Math Kangaroo USA
Problems and Solutions

Grades 3 & 4

Even Years
1998–2022

Editor in Chief

Agata Gazal
Chief Editorial Officer for Math Kangaroo USA
Billings, MT

Reviewers and Contributors

Joanna Matthiesen
Chief Executive Officer for Math Kangaroo USA
Granger, IN

Izabela Szpiech
Chief Financial Officer for Math Kangaroo USA
Chicago, IL

Kasia Nalaskowska
Chief Information Officer for Math Kangaroo USA
Aurora, IL

Magdalena Teodorowicz
Chief Design Officer for Math Kangaroo USA
Cordova, TN

Professor Andrzej Zarach, Ph.D.
Math Content Reviewer
East Stroudsburg University, East Stroudsburg, PA

Book Design

Jossea K. Rilea
Designer at LX Design Lab
Saratoga Springs, NY

We would like to give special thanks to other countless people who contributed to the problems and solutions for this book since 1998. Primarily, a big thank you to the Math Kangaroo question writers from all over the world that are part of the AKSF organization (www.aksf.org). Math Kangaroo solution writers also include Math Kangaroo USA competition organizers and Math Kangaroo Alumni. We would also like to thank the hundreds of educators who gave us feedback on the questions and solutions and finally the hundreds of thousands of students that took the Kangaroo challenge over the last two decades. Thank you all for your help in developing this book.

www.mathkangaroo.org

For additional copies of this book, please contact the publisher:
Math Kangaroo USA
info@mathkangaroo.org

ISBN: 979-8-9899883-0-3

Hop to the Top with This Official Math Kangaroo Book for Grades 3 and 4!

Our newest preparation book for the annual Math Kangaroo Competition promises to spark a child's interest in mathematics — it introduces motivated elementary school students to math riddles, puzzles, and creative logic questions in fun and enjoyable ways. Math Kangaroo books do not focus on teaching arithmetic; instead, they are powerful tools in enhancing students' abilities to analyze and solve complex problems while using logical reasoning.

Inside the book, 3rd and 4th grade students will find 312 entertaining problems presented during actual Math Kangaroo Competition even years, spanning 1998-2022, for a total of 13 tests, and their solutions. Each test consists of 24 questions divided into color-coded easy, medium, and difficult categories.

The Math Kangaroo question creation and selection process is intrinsically diverse, coming from the minds of top-university professors from over 100 countries worldwide. The best of the best questions, derived from group scrutiny, are selected each year at the Kangourou sans Frontières meeting. Out of thousands of submissions, only the best-crafted, most engaging, and age-appropriate questions are selected. After the selection process, solutions to the questions are written by a devoted team of Math Kangaroo USA faculty members, professors, and alumni.

This easy-to-use resource book will help students practice their math skills and also use math and logic as a tool for understanding the world around them.

We hope this book will be cherished by students who love mathematics, parents who like to study math with their children at home, and educators passionate about teaching unconventional and challenging math.

Enjoy this book and please let us know how you like it at www.mathkangaroo.org.

Joanna Matthiesen
President and CEO, Math Kangaroo USA

COLOR KEY

Each test has 24 questions with 3 levels of difficulty

GREEN	YELLOW	RED
Easy	**Medium**	**Difficult**
Q 1-8	**Q 9-16**	**Q 17-24**
3 Points	**4 Points**	**5 Points**

Contents

TABLE OF CONTENTS

Part I
Problems

1998

Three problems (8, 16, and 24) were added to the original 1998 test.

3 Points Each

1 At the zoo, Bob saw kangaroos for the first time. He noticed that each kangaroo had four legs, two ears, and one tail. For fun, he counted all the legs, ears, and tails, and got the number 63. How many kangaroos did he see?

(A) 6
(B) 7
(C) 9
(D) 10
(E) 12

---

2 John wanted to place parentheses in the expression 6 × 8 + 20 ÷ 4 – 2 in such a way as to get 58 as the result. Which of the ways shown below would give him this result?

(A) 6 × (8 + 20) ÷ 4 – 2
(B) (6 × 8 + 20 ÷ 4) – 2
(C) (6 × 8 + 20) ÷ 4 – 2
(D) 6 × 8 + 20 ÷ (4 – 2)
(E) 6 × (8 + 20 ÷ 4) – 2

---

3 How many triangles can you see in the picture?

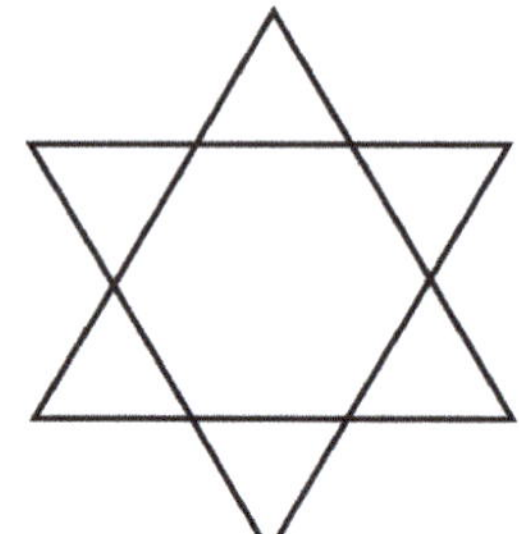

(A) 2
(B) 6
(C) 8
(D) 10
(E) 12

4 Mary lives in a tall building in apartment number 17. There are stores on the first floor. Above the stores, there are 3 apartments on each floor, numbered consecutively. On what floor does Mary live?

(A) 4th
(B) 5th
(C) 6th
(D) 7th
(E) 9th

---

5 What number do we need to place inside ☐ to make 12 × 12 × 12 = 6 × ☐ × 6 true?

(A) 12
(B) 24
(C) 48
(D) 72
(E) 60

---

6 How many three-digit numbers can you create using the digits 3, 0, and 7, and using each digit only once?

(A) 2
(B) 3
(C) 4
(D) 5
(E) 6

---

7 Out of how many blocks is this tower built?

(A) 20
(B) 22
(C) 25
(D) 28
(E) 30

8 I chose a number, then I multiplied it by 3 and added 18 to the result. I got 180. What was the number I chose?

(A) 52
(B) 54
(C) 50
(D) 48
(E) 62

4 Points Each

9 A kangaroo is traveling from START to FINISH using the paths shown in the picture. Each segment is marked with the time (in minutes) which the kangaroo needs to travel that segment. What is the shortest time needed for the kangaroo to reach FINISH?

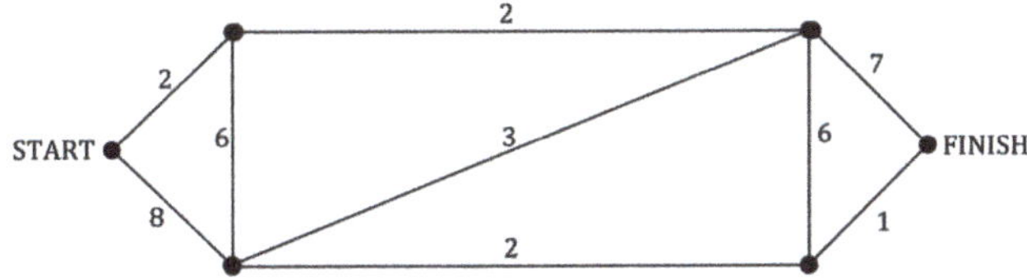

(A) 11 minutes
(B) 8 minutes
(C) 10 minutes
(D) 18 minutes
(E) 6 minutes

10 Joanna baked some cookies. She tried to divide them evenly, first between 2 plates, then between 3 plates, and finally between 4 plates. Each time she had one cookie left over. What is the smallest number of cookies Joanna could have baked?

(A) 9
(B) 10
(C) 11
(D) 12
(E) 13

11 I chose a certain number. I then subtracted 40 from it. Then I added 2000 and as a result I now have 3250. What number did I choose in the beginning?

(A) 2040
(B) 1960
(C) 1290
(D) 3210
(E) 1250

12 On Monday morning, a snail fell down a well which is 5 meters deep. During the day, it climbs up 2 meters, but during the night it slides down 1 meter. On what day of the week will the snail get out of the well?

(A) Tuesday
(B) Wednesday
(C) Thursday
(D) Friday
(E) Monday

13 Adam cut five identical square sheets of paper into two pieces. From which of the five pieces below was the piece marked with Z cut?

Z

(A) 

(B) 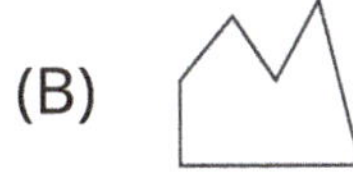

(C) 

(D) 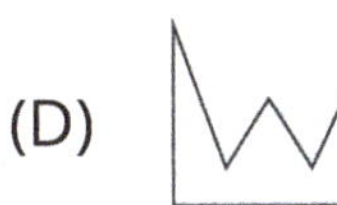

(E) 

14 There were 31 runners competing in a race. The number of runners who finished before John is four times smaller than the number of runners who finished later than John. In what place did John finish?

(A) 6
(B) 7
(C) 8
(D) 20
(E) 21

---

15 Half a loaf of bread costs 6 pence more than one-fourth of a loaf of bread. How many pence does a whole loaf of bread cost? (Note: A pence is an English coin.)

(A) 6
(B) 12
(C) 18
(D) 24
(E) 30

---

16 Little Kangaroo's clock fell from the wall. The Kangaroo does not know which side should go up. Can you help her and tell what the time is?

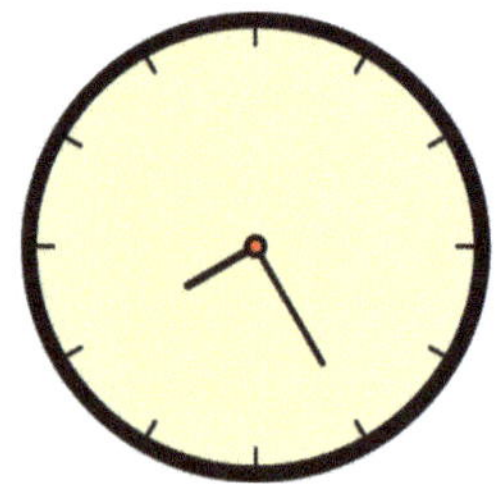

(A) 11:45
(B) 9:00
(C) 8:25
(D) 6:35
(E) 3:00

5 Points Each

17 Write the numbers into the blank boxes in the pyramid according to the pattern shown below.

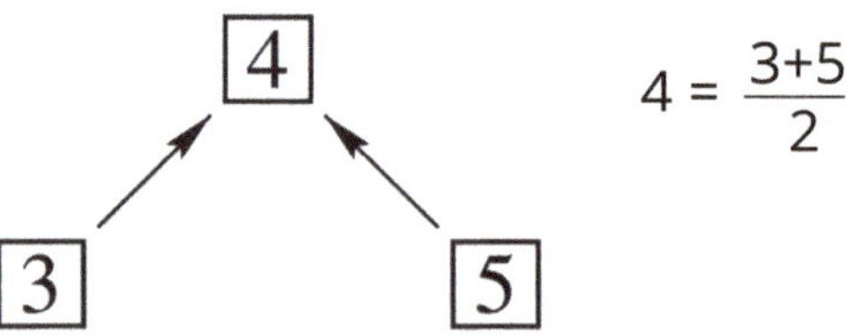

What number is at the top of the pyramid?

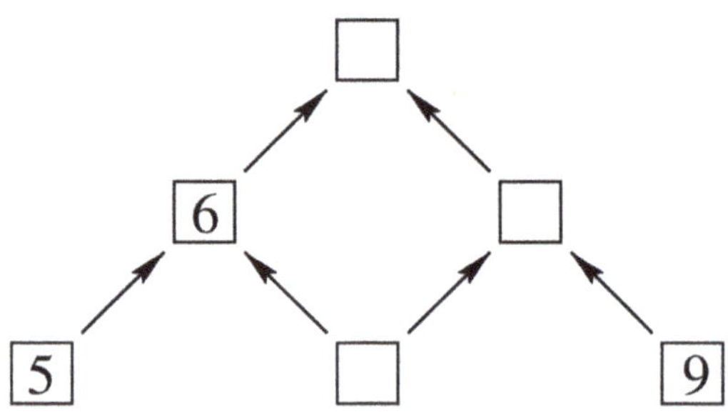

(A) 5
(B) 7
(C) 8
(D) 9
(E) 12

---

18 There are 15 balls in a box: white balls, red balls, and black balls. The number of white balls is 7 times greater than the number of red balls. How many black balls are there in the box?

(A) 1
(B) 3
(C) 5
(D) 7
(E) 9

19 Paul was going to buy 4 servings of ice cream, but he was 80 cents short. So, he bought 3 servings and had 30 cents left. What was the price of one serving of ice cream?

(A) 70 cents
(B) 80 cents
(C) 90 cents
(D) 1 dollar
(E) 1 dollar and 10 cents

---

20 We are making a square "chessboard" using matches that are 5 centimeters long. One side of the chessboard will be 1 meter long.

The picture shows the upper left-hand corner of the board.

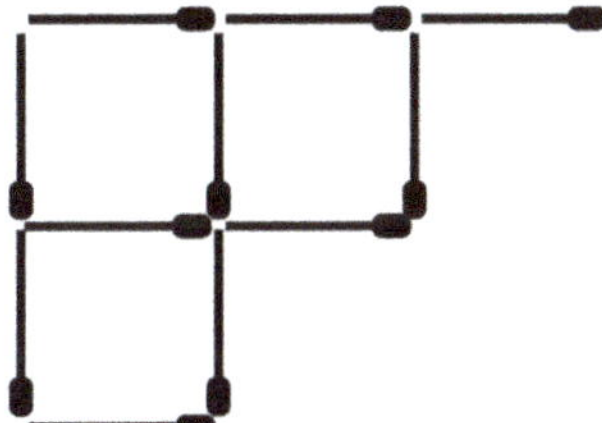

How many matches will we use?
(Note: 1 m = 100 cm.)

(A) 400
(B) 480
(C) 640
(D) 840
(E) 960

---

21 How many three-digit numbers are there that have the sum of their digits equal to 5?

(For example, 122 is such a number, because 1 + 2 + 2 = 5.)

(A) 10
(B) 15
(C) 20
(D) 25
(E) 30

22 Ann is 3 years older than Barb and 2 years younger than Cali. Dorothy is 1 year younger than Barb. How much older is Cali than Dorothy?

(A) 5 years
(B) 6 years
(C) 4 years
(D) 2 years
(E) They are the same age.

---

23 In a certain soccer tournament, the winning team gets 3 points, the losing team gets 0 points, and in the case of a tie both teams get 1 point each. My team played 31 games and received 64 points. 7 of the games were ties. How many games did my team lose?

(A) 0
(B) 5
(C) 19
(D) 21
(E) 24

---

24 The number 12345678910111213141516... is made by writing consecutive natural numbers in order starting from 1. Which natural number does the digit in the 180th place belong to?

(A) 92
(B) 93
(C) 94
(D) 95
(E) 96

2000

3 Points Each

1 A birthday candle stays lit for 15 minutes. For how long will 10 birthday candles stay lit if they are lit at the same time and no one blows them out?

(A) 1.5 minutes
(B) 15 minutes
(C) 150 minutes
(D) 1.5 hours
(E) 15 hours

---

2 The little kangaroo was sick. Dr. Ohpain prescribed 3 pills for him to take one at a time every twenty minutes. How many minutes after taking the first pill will the little kangaroo take the last pill?

(A) 20
(B) 30
(C) 40
(D) 50
(E) 60

---

3 For which of the following numbers is the product of the digits greater than the sum of the digits?

(A) 112
(B) 209
(C) 312
(D) 222
(E) 211

4 Gavel lives on the second floor, and Pavel lives in the same building but has to walk up twice as many stairs as Gavel. There are no stairs to the entrance of the building. On which floor does Pavel live?

(A) on the 2nd floor
(B) on the 3rd floor
(C) on the 4th floor
(D) on the 5th floor
(E) on the 6th floor

---

5 Four candy bars and three lollipops cost $4.50. One candy bar costs 90 cents. How much does one lollipop cost?

(A) 20 cents
(B) 30 cents
(C) 40 cents
(D) 50 cents
(E) 60 cents

---

6 One tour bus can seat no more than 55 people. What is the smallest number of buses needed to seat 160 people?

(A) 1
(B) 2
(C) 3
(D) 4
(E) 5

---

7 A person needs 12 minutes to walk around a square plaza. How much time will it take for the same person to walk at the same pace around a plaza that has an area that is four times greater?

(A) 48 minutes
(B) 24 minutes
(C) 30 minutes
(D) 20 minutes
(E) 36 minutes

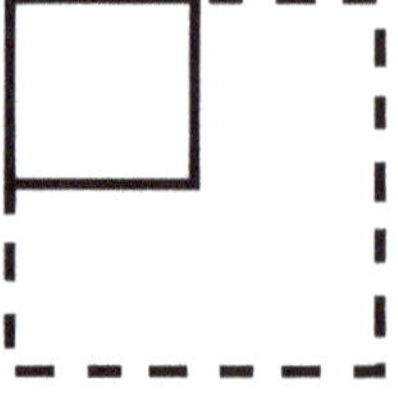

8 It is 120 km from Zakopane to the Krakow airport. Buses leave from Zakopane for the airport at 30 minutes past every hour. The buses drive with an average speed of 60 km per hour. A group of "Kangaroos," members of a math camp in Zakopane, is supposed to arrive at the airport at 11:30. What is the latest time their bus had to leave Zakopane to get them to the airport on time?

(A) 7:30
(B) 8:30
(C) 9:30
(D) 10:30
(E) 11:30

4 Points Each

9 During the time that Kathy eats two bowls of ice cream, Betty eats three bowls of ice cream. The two girls ate 10 bowls of ice cream in one hour. How many bowls of ice cream did Kathy eat?

(A) 3
(B) 4
(C) 5
(D) 6
(E) 7

10 Which four digits need to be removed from the number 4921508 to get the smallest possible three-digit number?

(A) 4, 9, 2, 1
(B) 4, 2, 1, 0
(C) 1, 5, 0, 8
(D) 4, 9, 2, 5
(E) 4, 9, 5, 8

11 In each of two baskets there were 12 apples. Aria took a certain number of apples from the first basket. From the second basket, Zoe took a number of apples equal to the number of apples remaining in the first basket. How many apples were left in both baskets altogether?

(A) 6
(B) 12
(C) 18
(D) 20
(E) 24

12 Students walked to the museum in rows of three. Adam, Bart, and Carl noticed that they were in the seventh row from the front and in the fifth row from the back. How many students went to the museum?

(A) 12
(B) 24
(C) 30
(D) 33
(E) 36

13 Fourteen cats took part in a cat play. Some of them played moms and some of them played their kids. Every mom in this play had at least two kids. What is the greatest possible number of cat-moms in the play?

(A) 3
(B) 4
(C) 5
(D) 6
(E) 7

14 The first and the second scales are balanced (see the picture). How many plums do you need to place on the left side of the third scale to keep it in balance?

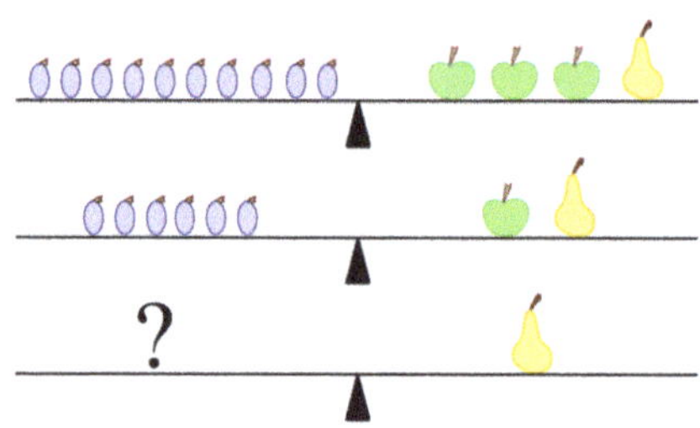

(A) 2
(B) 3
(C) 4
(D) 5
(E) 6

15 Each of the five neighbors owns a rectangular plot of land with the same area. The parts of the land with flowers growing on them are fenced in (solid line in the pictures). Who has the longest fence?

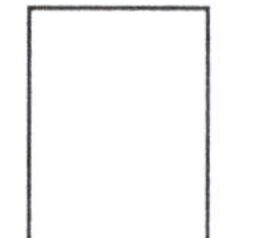
(A) Mr. Adam

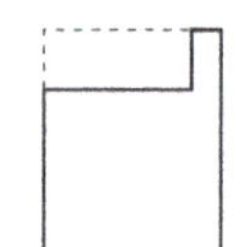
(B) Mr. John

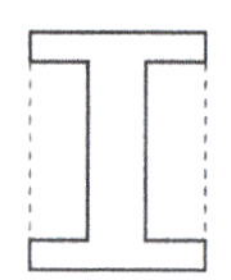
(C) Mr. Jack

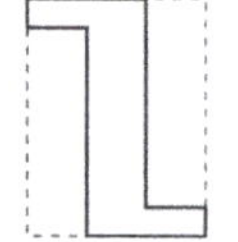
(D) Mr. Peter

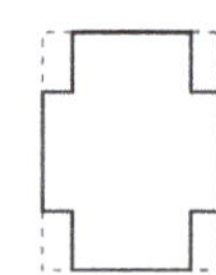
(E) Mr. Mark

16 For his birthday Patrick got a box with some identical cube blocks. He used all of them to make two projects (see the picture). All the blocks together weigh 900 grams. The project on the left weighs 300 grams and the picture shows all the blocks it is made of. How many blocks in the figure on the right are not shown?

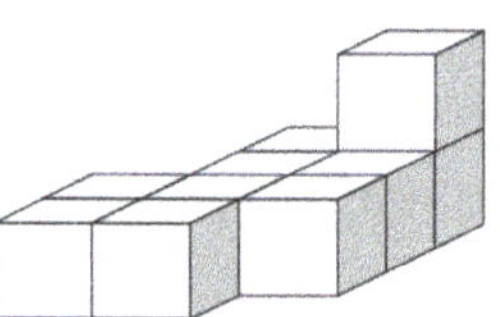
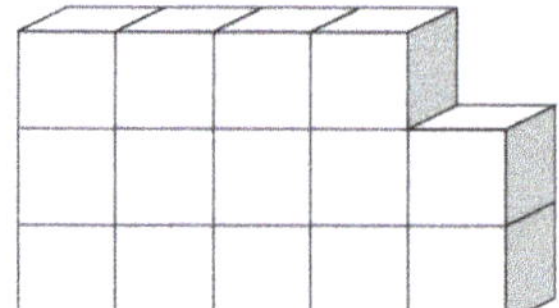

(A) 4
(B) 5
(C) 6
(D) 7
(E) 8

5 Points Each

17 Altogether, 6 hens eat 8 cups of grain in 3 days. How many cups of grain will 3 hens eat in 9 days?

(A) 10
(B) 12
(C) 14
(D) 16
(E) 9

18 Adrianna's birthday present is placed in a box with dimensions of 10 cm × 10 cm × 30 cm and wrapped with a ribbon as shown in the picture. What is the length of the ribbon?

(A) 200 cm
(B) 240 cm
(C) 260 cm
(D) 300 cm
(E) 250 cm

19 Three kangaroos were born consecutively every 4 years. Right now, the oldest kangaroo is 5 times as old as the youngest one. How old is the youngest kangaroo?

(A) 10
(B) 8
(C) 6
(D) 4
(E) 2

---

20 When Mary was leaving home between 8 and 9 o'clock in the morning, she noticed that the hour hand and the minute hand on her watch were overlapping. When she returned home between 2 and 3 o'clock in the afternoon, the hour hand and the minute hand formed a straight line (see the picture). How long was Mary away from home?

 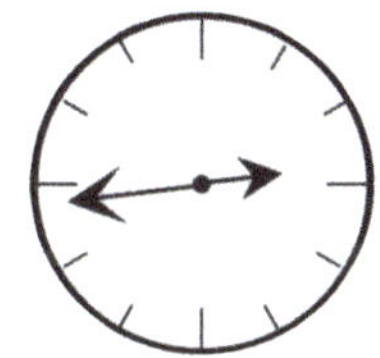

(A) 5 hours
(B) 5 and a half hours
(C) 6 hours
(D) 6 and a half hours
(E) 7 hours

---

21 Find the number which has the following property: If we add together this number and half of this number, we will get a number which is 3 less than twice the original number.

(A) 2
(B) 4
(C) 6
(D) 8
(E) 10

---

22 Three identical dice are placed one on top of the other (see the picture). The sides which touch each other have the same numbers of dots. What is the number of dots on the bottom of the lowest die?

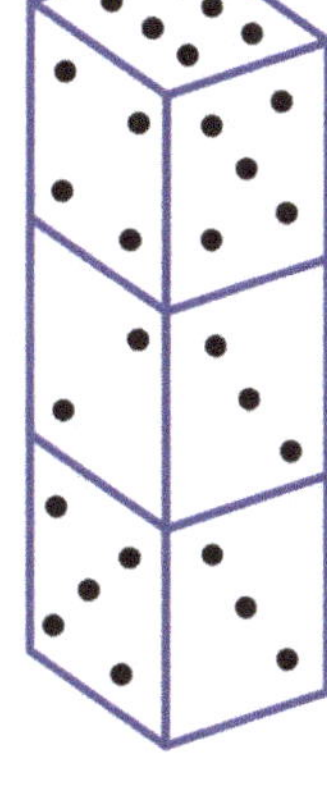

(A) 1
(B) 2
(C) 3
(D) 5
(E) 6

---

23 Pete wanted to draw the picture of a kangaroo shown, without lifting his pencil from the paper and without going over the same line twice. At what point should he start (see the picture)?

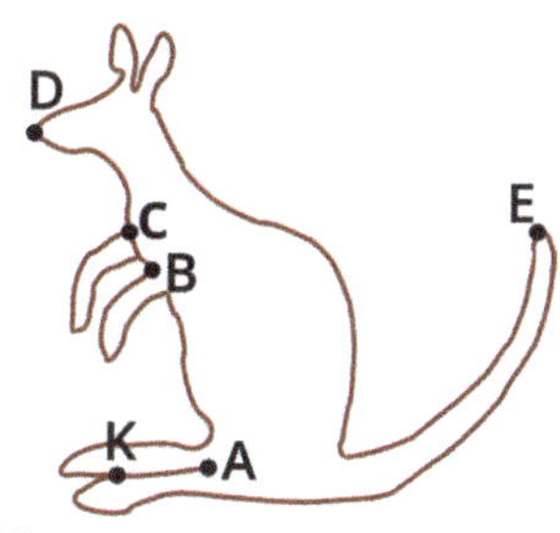

(A) A
(B) B or C
(C) D or E
(D) K
(E) There is no such point; this is impossible.

---

24 A magical ball falling to the ground bounces twice as high as the height from which it was dropped. From what height was the ball dropped if it reached the height of 320 cm after the second bounce?

(A) 80 cm
(B) 160 cm
(C) 320 cm
(D) 640 cm
(E) 1280 cm

2002

3 Points Each

1 Which of the squares below should be put into the picture to get the symbol of our competition?

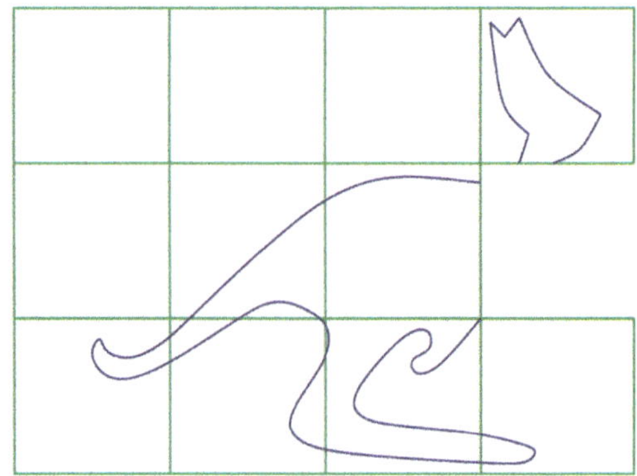

(A)

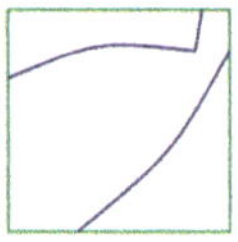

(B)

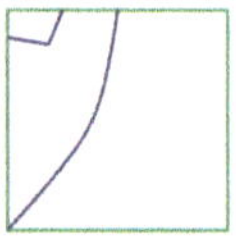

(C)

(D)

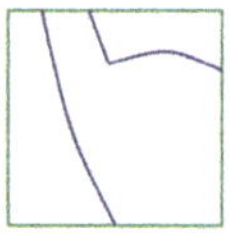

(E)

2 After we simplify
2 + 2 – 2 + 2 – 2 + 2 – 2 + 2 – 2 + 2
the result will be:

(A) 0
(B) 2
(C) 4
(D) 12
(E) 20

3 Andrzej received three cars, four balls, three teddy bears, ten pens, two chocolate bars, and a book for his birthday.
How many items did he get in all?

(A) 15
(B) 17
(C) 20
(D) 23
(E) 27

4 A square was divided into pieces (see the picture). Which of the pieces below does not occur in this divided square?

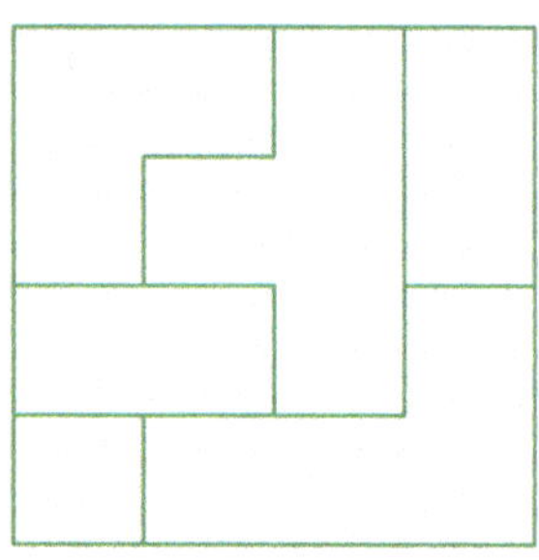

(A)

(B)

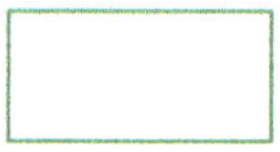

(C)

(D)

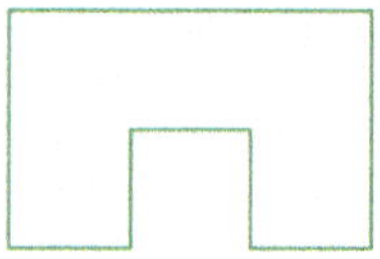

(E)

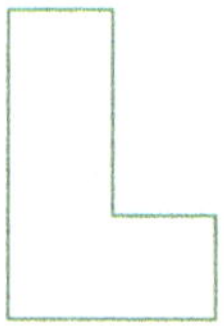

5 Gina, Lena, Suzie, and Joanna have their birthdays on March 1st, May 17th, July 20th, and March 20th (not necessarily in that order). Lena and Suzie were born in the same month. Gina and Suzie were born on the same day of the month. Which of the girls was born on May 17th?

(A) Gina
(B) Lena
(C) Suzie
(D) Joanna
(E) This cannot be determined from the information given.

---

6 The human heart beats an average of 70 times per minute. On average how many times does it beat during one hour?

(A) 42,000
(B) 7,000
(C) 4,200
(D) 700
(E) 420

---

7 Quadrilateral *ABCD* is a square and its side is 10 cm long. Quadrilateral *ATMD* is a rectangle and its shorter side is 3 cm. What is the difference between the sum of the lengths of all the sides of the square and the sum of the lengths of all the sides of the rectangle *ATMD*?

(A) 14 cm
(B) 10 cm
(C) 7 cm
(D) 6 cm
(E) 4 cm

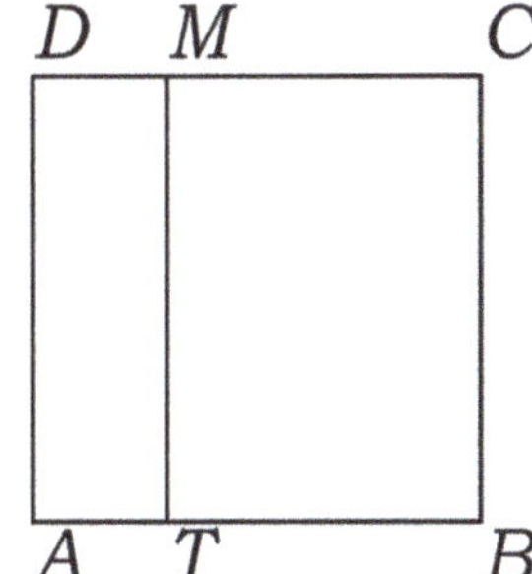

8 Which of the figures below couldn't be made by folding a rectangular sheet just once (see the picture)?

(A) 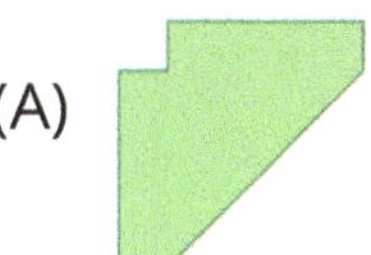

(B) 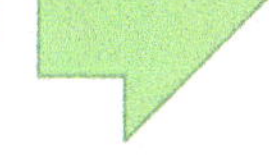

(C) 

(D) 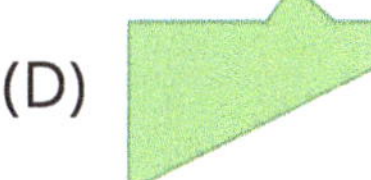

(E) 

4 Points Each

9 The houses on the street where John lives are numbered from 1 to 24. How many times does the digit 2 appear in the numbering of these houses?

(A) 2
(B) 4
(C) 8
(D) 16
(E) 32

---

10 There are six identical oranges on one side of the scale and two identical melons on the other side of the scale. After we add one melon to the side with the oranges, the scale will be balanced. How many oranges weigh as much as one melon?

(A) 2
(B) 3
(C) 4
(D) 5
(E) 6

11 The picture below is a sketch of a castle. Which of the lines below is not part of the sketch?

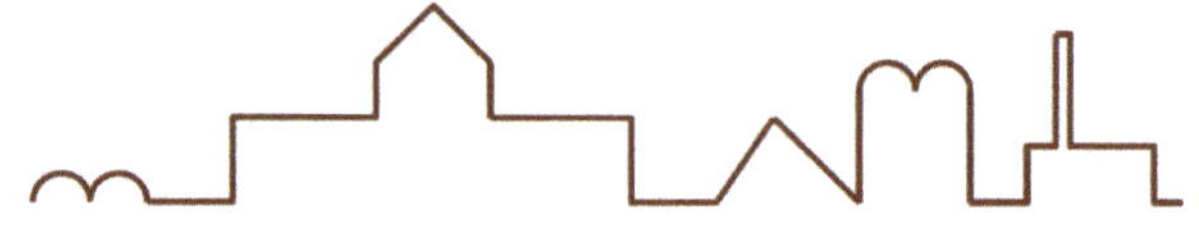

(A)

(B)

(C)

(D)

(E)

12 We add 17 to the smallest two-digit number and then we divide the sum by the largest one digit number. What is the result?

(A) 3
(B) 6
(C) 9
(D) 11
(E) 27

13 In a certain ancient country the numbers one, ten, and sixty were expressed with the following symbols:

1 10 60

People were writing down other numbers using these symbols. For example, the number 22 was written as:

Which of the following notations represents the number 124?

(A)

(B)

(C)

(D)

(E)

14 The face of a clock was divided into four parts. The sums of the numbers in each of those parts are consecutive numbers. Which of the following pictures satisfies this rule?

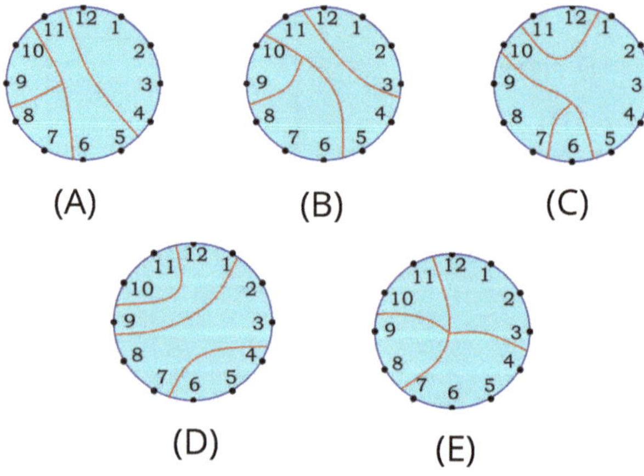

(A) (B) (C)

(D) (E)

15 Klara and Zoe had 60 matches altogether. Klara took as many matches as she needed to build a triangle with each side 6 matches long. Zoe used the remaining matches to build a rectangle which had one side equal to 6 matches. How many matches did she use to make one longer side of this rectangle?

(A) 9
(B) 12
(C) 15
(D) 18
(E) 30

---

16 Three kangaroos, Miki, Niki, and Oki, participated in a competition. Jumping at the same speed, they jumped along the lines you can see in the picture. Only one of the sentences A, B, C, D, and E listed below is true. Which one is it?

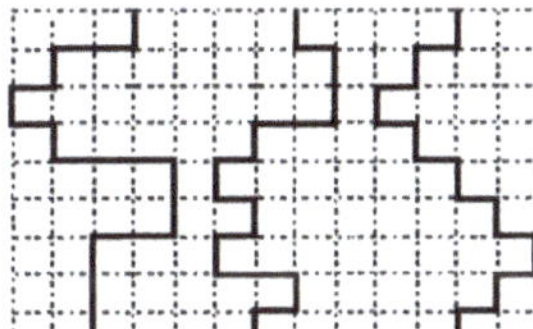

Miki Niki Oki

(A) Miki and Oki finished at the same time.
(B) Niki finished first.
(C) Oki finished last.
(D) All the kangaroos finished at the same time.
(E) Miki and Niki finished at the same time.

5 Points Each

17 Each boy, Mike, Nate, Oliver, and Paul, has exactly one of the following animals: a cat, a dog, a goldfish, and a canary. Nate has a pet with fur. Oliver has a pet with four legs. Paul has a bird, and Mike and Nate don't like cats. Which of the following sentences is not true?

(A) Oliver has a dog.
(B) Paul has a canary.
(C) Mike has a goldfish.
(D) Oliver has a cat.
(E) Nate has a dog.

---

18 Mary leaves her house at 6:55 and arrives at school at 7:32. Zoe needs 12 minutes less than Mary to get to school. Yesterday Zoe showed up at school at 7:45. What time did she leave her house?

(A) at 7:07
(B) at 7:20
(C) at 7:25
(D) at 7:30
(E) at 7:33

19 Robert had a certain number of identical cubes. He made a tunnel using half of his blocks (see Picture 1). With some of the remaining cubes he formed a pyramid (see Picture 2). How many blocks were left after making both of these structures?

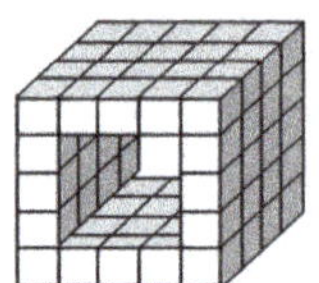
Picture 1

Picture 2

(A) 34
(B) 28
(C) 22
(D) 18
(E) 15

---

20 The daughter is 3 years old, and her mom is 28 years older than the daughter. How many years later will the mom be three times as old as her daughter?

(A) 9
(B) 12
(C) 10
(D) 1
(E) 11

---

21 A conductor wanted to make a trio consisting of a violinist, a pianist, and a drummer. He had to choose one of two violinists, one of two pianists, and one of two drummers. He decided to try each of the possible trios. How many attempts did he have to make?

(A) 3
(B) 4
(C) 8
(D) 24
(E) 25

---

22 One medal can be cut out from a golden square plate. If four medals are made from four plates, the remaining parts of those four plates can be used to make one more plate. What is the largest number of medals that can be made when 16 plates are used?

(A) 17
(B) 19
(C) 20
(D) 21
(E) 32

---

23 Twenty eight students from the fourth grade competed in a math competition. Each student earned a different number of points. The number of students who received more points than Tom is two times smaller than the number of students who had fewer points than Tom. In which position did Tom finish that competition?

(A) 6th
(B) 7th
(C) 8th
(D) 9th
(E) 10th

---

24 An odometer in a car shows the number 187569, which is the number of kilometers that have been traveled. This number consists of all different digits. After traveling how many more kilometers will the odometer show a number consisting of all different digits again?

(A) after 777 km
(B) after 12,431 km
(C) after 431 km
(D) after 21 km
(E) after 11 km

2004

3 Points Each

1 2001+ 2002 + 2003 + 2004 + 2005 =

(A) 1,015
(B) 5,010
(C) 10,150
(D) 11,005
(E) 10,015

---

2 Mark was 4 years old when his sister was born. Today he blew out all 9 candles on his birthday cake. What is the difference between Mark's age and his sister's age today?

(A) 4 years
(B) 5 years
(C) 9 years
(D) 13 years
(E) 14 years

---

3 The picture below shows a road from town A to town B (indicated by a solid line) and a detour (marked by a dashed line) caused by renovations of the section CD. How many kilometers longer is the road from town A to town B because of the detour?

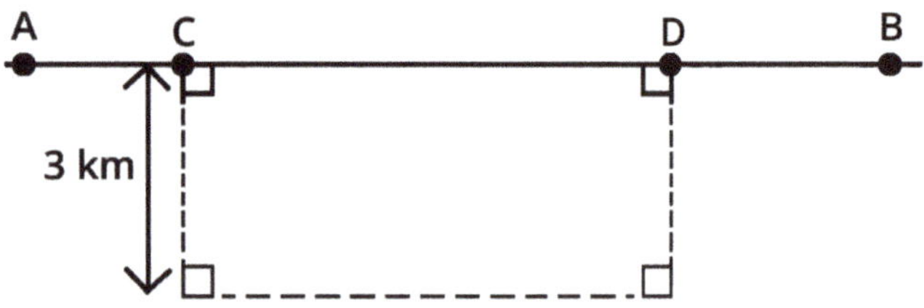

(A) 3 km
(B) 5 km
(C) 6 km
(D) 10 km
(E) This cannot be calculated.

4 Which of the results below is not identical to the difference 671 – 389?

(A) 771 – 489
(B) 681 – 399
(C) 669 – 391
(D) 1871 – 1589
(E) 600 – 318

---

5 There were some birds sitting on the telegraph wire. Then, 5 of them flew away and after some time 3 birds came back. At that time there were 12 birds sitting on the wire. How many birds were there at the very beginning?

(A) 8
(B) 9
(C) 10
(D) 12
(E) 14

---

6 Which numbers are inside the rectangle and inside the circle but not inside the triangle as well?

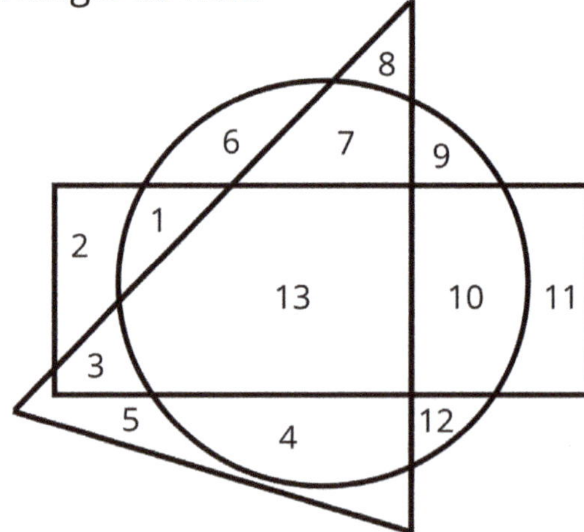

(A) 5 and 11
(B) 1 and 10
(C) 13
(D) 3 and 9
(E) 6, 7, and 4

7 The buildings on Color Street are numbered from 1 to 5 (see the picture).

Each building is colored with one of the following colors: blue, red, yellow, pink, and green. We know that:

- The red building is next only to the blue building.
- The blue building is between the red building and the green building.

What is the color of the building with the number 3?

(A) blue
(B) red
(C) yellow
(D) pink
(E) green

---

8 How many white squares need to be shaded in the picture so that the number of shaded squares is equal to exactly half of the number of white squares?

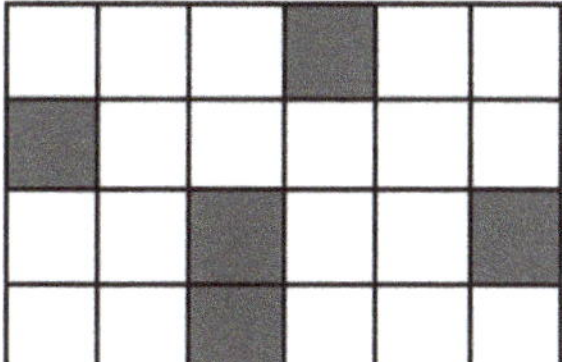

(A) 2
(B) 3
(C) 4
(D) 6
(E) It is impossible to calculate.

4 Points Each

9 Five identical rectangular plastic sheets were divided into white and black squares. Which of the sheets from A to E has to be covered with the sheet in picture 1 in order to get a completely black rectangle?

Picture 1

(A) 

(B) 

(C) 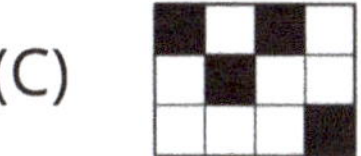

(D) 

(E) 

---

10 The scales in the pictures have been balanced. There are pencils and a pen on the two sets of scales. What is the weight of the pen in grams?

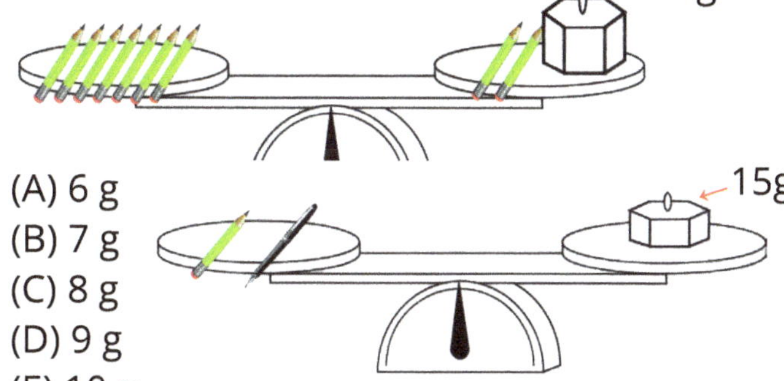

(A) 6 g
(B) 7 g
(C) 8 g
(D) 9 g
(E) 10 g

11 I noticed four clocks on the wall. Only one of them shows the correct time. One of them is 20 minutes ahead, another is 20 minutes late, and yet another one is broken. What time is it right now?

(A) 4:45
(B) 5:05
(C) 5:25
(D) 5:40
(E) 12:00

---

12 Ella brought a basket of apples and oranges to a birthday party. Guests ate half of the apples and a third of the oranges. What remained in the basket was:

(A) half of all the fruit.
(B) more than half of all the fruit.
(C) less than half of all the fruit.
(D) a third of all the fruit.
(E) less than a third of all the fruit.

---

13 Angie divided a certain number by 10 instead of multiplying it by 10. As a result she got 600. What would the result have been if she hadn't made this mistake?

(A) 6
(B) 60
(C) 600
(D) 6,000
(E) 60,000

14 Kathy found a book which was missing some pages. When she opened the book, she saw the number 24 on the left side and the number 45 on the right side. How many sheets of paper were missing between these two pages?

(A) 9
(B) 10
(C) 11
(D) 20
(E) 21

---

15 Eva is 52 days older than her friend Andrea. Eva had her birthday on Tuesday in March this year. On which day of the week will Andrea celebrate her birthday this year?

(A) Monday
(B) Tuesday
(C) Wednesday
(D) Thursday
(E) Friday

---

16 Numbers were placed in a square table in such a way that the sum of the numbers in the first row is equal to 3, the sum of the numbers in the second row is equal to 8, and the sum of the numbers in the first column is equal to 4. What is the sum of the numbers in the second column?

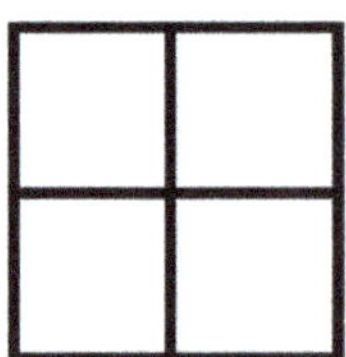

(A) 4
(B) 6
(C) 7
(D) 8
(E) 11

5 Points Each

17 A certain cube is painted with three colors so that every side of this cube is one of the colors and opposite sides are the same color. From which of the patterns below can this kind of cube be made?

(A) 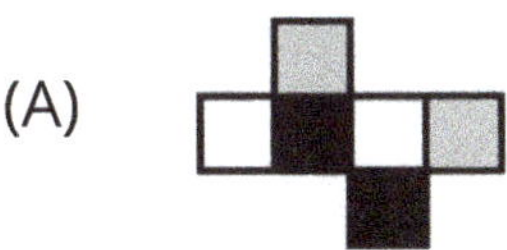

(B) 

(C) 

(D) 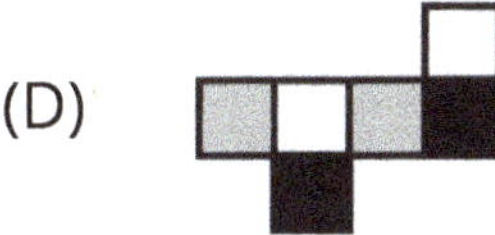

(E) 

---

18 Four square tiles were arranged in the way shown in the picture. The lengths of the sides of two tiles are indicated. What is the length of the side of the largest tile?

(A) 24
(B) 56
(C) 64
(D) 81
(E) 100

16

40

---

19 Girls and boys from Maria and Matt's class formed a line. There are 16 students on Maria's right, and Matt is among them. There are 14 students on Matt's left, and Maria is among them. There are 7 students between Maria and Matt. How many students are there in this class?

(A) 37
(B) 30
(C) 23
(D) 22
(E) 16

---

20 The sum of the digits of the 10-digit number is 9. What is the product of the digits of this number?

(A) 0
(B) 1
(C) 45
(D) $9 \times 8 \times 7 \times ... \times 2 \times 1$
(E) 10

---

21 The big cube was formed out of 125 small cubes, each of which is either black or white (see the picture). Any two adjacent cubes have different colors. The vertices of the big cube are black. How many white cubes does the big cube contain?

(A) 62
(B) 63
(C) 64
(D) 65
(E) 68

22 A lottery ticket costs 4 dollars. Three boys, Paul, Peter, and Robert, collected money and bought two tickets. Paul gave 1 dollar, Peter gave 3 dollars, and Robert gave 4 dollars. One of the tickets they bought won 1000 dollars. Boys shared the award fairly, that is, proportionally to their contributions. How much did Peter receive?

(A) \$300
(B) \$375
(C) \$250
(D) \$750
(E) \$425

---

23 In three soccer games Daniel's team scored three goals and had one goal scored against them. For every game won the team gets 3 points, for a tie it gets 1 point, and for a game lost it gets 0 points. It is certain that the number of points the team earned in those three games was not equal to which of the following numbers?

(A) 7
(B) 6
(C) 5
(D) 4
(E) 3

24 Natural numbers were placed in a table. In each white cell of the table, the products of two numbers from the gray sections—one from above and one from the left—was placed (for example: 42 = 7 × 6). Some of these products are represented by letters. Which two letters represent the same number?

(A) L and M
(B) T and N
(C) R and P
(D) K and P
(E) M and S

×				7
	J	K	L	56
	M	36	8	N
	T	27	6	P
6	18	R	S	42

2006

3 Points Each

1 During a summer math camp in the city of Zakopane in Poland, the students take part in a trip to the top of Mount Giewont. It takes 3 hours to get to the top of the mountain. They stay at the top of the mountain for half an hour. Afterwards, it takes two and a half hours to come down the mountain. What time in the morning at the latest does the trip need to start so that everybody is back at the camp for a meal at 3 p.m.?

(A) 8:00
(B) 8:30
(C) 9:00
(D) 9:30
(E) 10:00

2 What is the value of this expression: $2 \times 0 \times 0 \times 6 + 2006$?

(A) 0
(B) 2006
(C) 2014
(D) 2018
(E) 4012

3 How many cubes have been removed from the first block to obtain the second block?

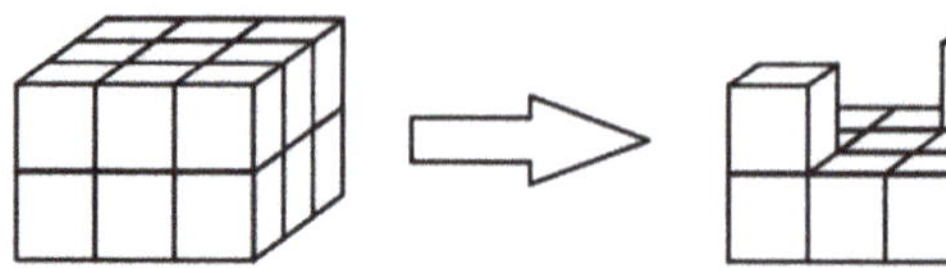

(A) 4
(B) 5
(C) 6
(D) 7
(E) 9

4 Katie's birthday was yesterday. It will be Thursday tomorrow. What day was Katie's birthday?

(A) Tuesday
(B) Wednesday
(C) Thursday
(D) Saturday
(E) Monday

5 John is playing with darts. He brings back all the darts after he has thrown them, and for each time he hits the bullseye, he gains two additional darts. At the beginning he has 10 darts and at the end he has 20. How many times did he hit the bullseye?

(A) 6
(B) 8
(C) 10
(D) 5
(E) 4

6 A kangaroo enters the building as shown in the picture. He only passes through triangular rooms. Where does he leave the building?

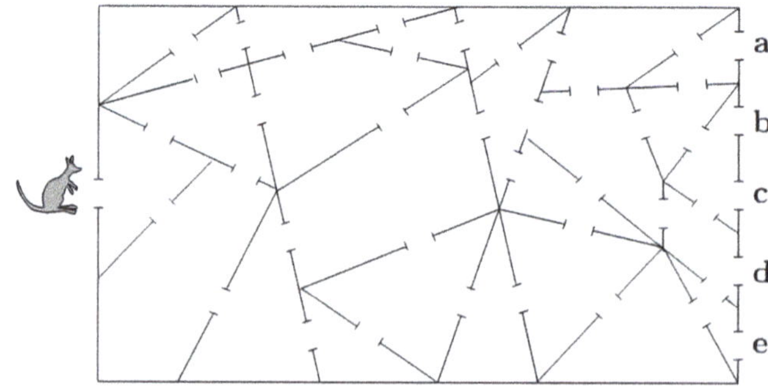

(A) a
(B) b
(C) c
(D) d
(E) e

7 Four people can sit around a certain kind of a square table. For the school party the students put together 7 such square tables in order to make one long rectangular table. How many people can sit at this long table now?

(A) 14
(B) 16
(C) 21
(D) 24
(E) 28

8 In his wallet, Stan has one 5-dollar bill, one 2-dollar bill, and one 1-dollar bill. Which of the following amounts can Stan not make using the bills that he has?

(A) $3
(B) $4
(C) $6
(D) $7
(E) $8

4 Points Each

9 On one side of Long Street the houses are numbered with consecutive odd numbers from 1 to 19. On the other side of that street, the houses are numbered with consecutive even numbers from 2 to 14. How many houses are there on Long Street?

(A) 8
(B) 16
(C) 17
(D) 18
(E) 33

10 From which of the figures below can the figure outlined in bold be cut out?

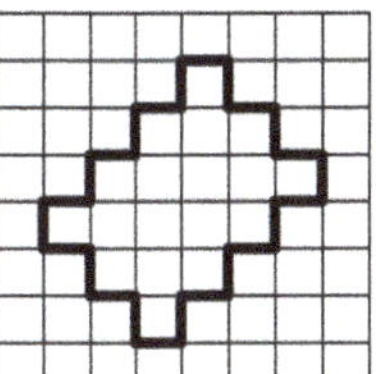

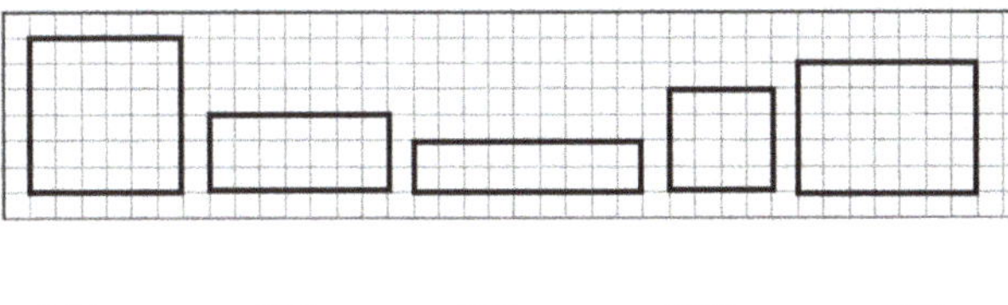

(A) (B) (C) (D) (E)

11 The picture below shows bus routes and ticket prices between 6 towns. What is the least amount of money needed to pay for the tickets to get from town A to town B?

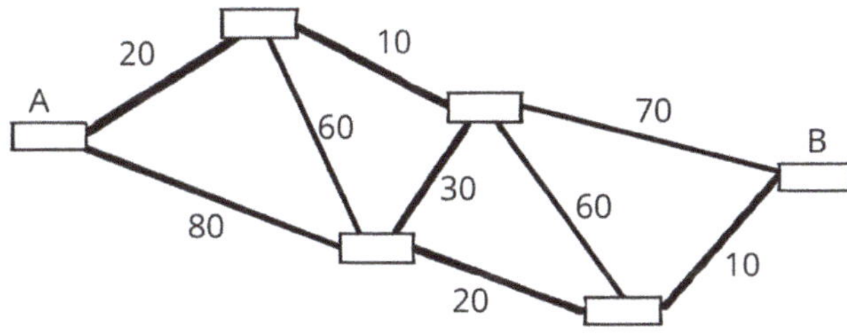

(A) 90
(B) 100
(C) 110
(D) 180
(E) 200

12 What is the smallest number we can get by arranging the six cards with numbers shown in the picture in one row, one after another?

(A) 1234567890
(B) 1023456789
(C) 3097568241
(D) 2309415687
(E) 2309415678

13 Six weights, weighing 1 pound, 2 pounds, 3 pounds, 4 pounds, 5 pounds, and 6 pounds, were placed into three boxes, two weights in each box. The weights in the first box weigh 9 pounds together, and those in the second box weigh 8 pounds. Which weights are in the third box?

(A) 5 and 2
(B) 6 and 1
(C) 3 and 1
(D) 4 and 2
(E) 4 and 3

---

14 Four routes are drawn between two points. Which route is the shortest?

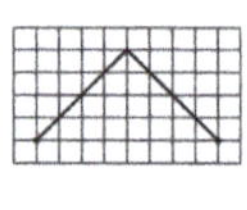
(A)

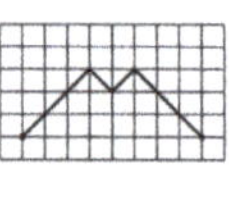
(B)

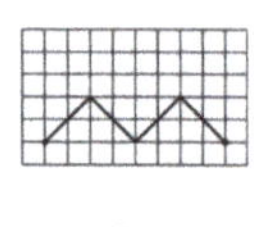
(C)

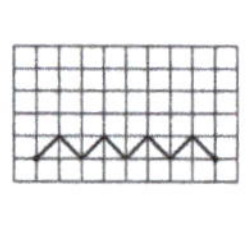
(D)

(E) All are equal.

---

15 Numbers are written on a "number flower." Mary picked off all the petals with numbers which give a remainder 2 when divided by 6. What is the sum of the numbers on the petals that Mary picked off?

(A) 46
(B) 66
(C) 84
(D) 86
(E) 114

16 You can move or rotate any of the shapes of the puzzle, but you cannot flip them over. Which of the shapes below does not appear in the puzzle?

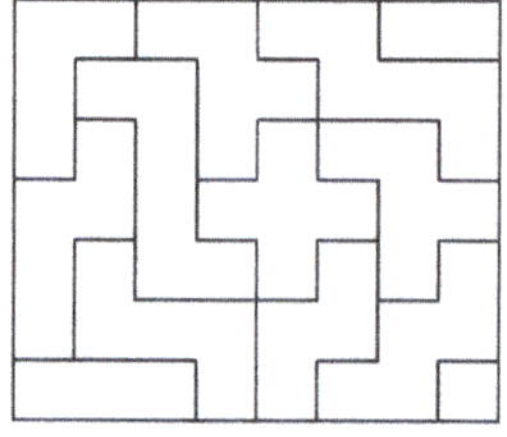

(A) 

(B) 

(C) 

(D) 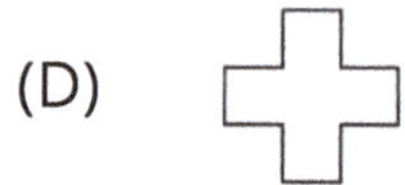

(E) 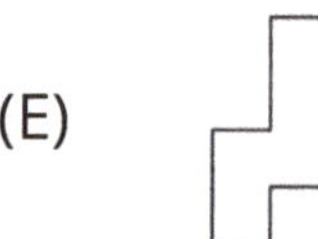

5 Points Each

17 Four crows are sitting on a fence. Their names are Dana, Hanna, Lena, and Bennie. Dana sits exactly in the middle between Hanna and Lena. The distance between Hanna and Dana is the same as the distance between Lena and Bennie. Dana sits 4 feet away from Bennie. How far away is Hanna sitting from Bennie?

(A) 5 feet
(B) 6 feet
(C) 7 feet
(D) 8 feet
(E) 9 feet

18 Johnny is building a house out of cards. In the picture, one-story, two-story, and three-story houses are shown. How many cards does Johnny need to build a 4-story house?

(A) 23
(B) 24
(C) 25
(D) 26
(E) 27

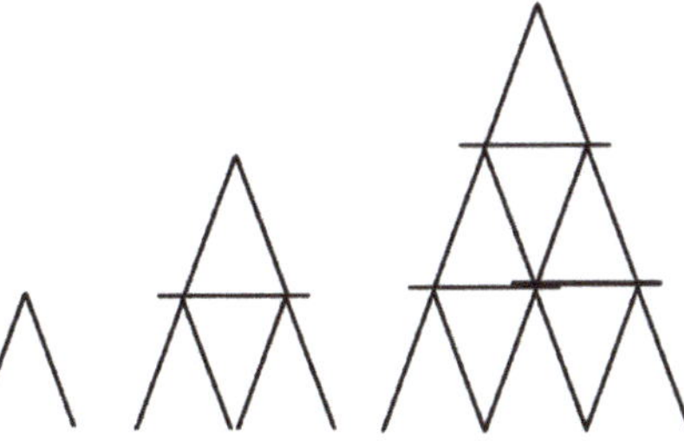

19 The structure shown in the picture is made by gluing together the sides of 10 cubes. Roman painted the entire structure, including the bottom. How many faces of the cubes did he paint?

(A) 18
(B) 24
(C) 30
(D) 36
(E) 42

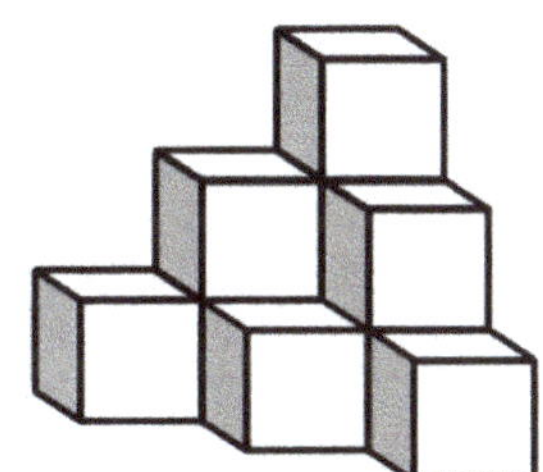

20 Irena, Ann, Kate, Olga, and Elena live in the same two-story house. Two of the girls live on the first floor; three of them live on the second floor. Olga lives on a different floor than Kate and Elena. Ann lives on a different floor than Irena and Kate. Who lives on the first floor?

(A) Kate and Elena
(B) Irena and Elena
(C) Irena and Olga
(D) Irena and Kate
(E) Ann and Olga

21 In the expression:
2002 □ 2003 □ 2004 □ 2005 □ 2006,
either "+" or "–" can be written in the place of each □. Which result is impossible?

(A) 1998
(B) 2001
(C) 2002
(D) 2004
(E) 2006

22 One year in March, there were 5 Mondays. Which day of the week below could not appear in this month five times as well?

(A) Saturday
(B) Sunday
(C) Tuesday
(D) Wednesday
(E) Thursday

23 In each of the nine cells of the square, we need to write one of the digits 1, 2, or 3. We need to do this in such a way that each of the digits 1, 2, and 3 will be written in each horizontal row and in each vertical column. If we start with 1 in the upper left cell, in how many different ways can the square be filled?

(A) 2
(B) 3
(C) 4
(D) 5
(E) 8

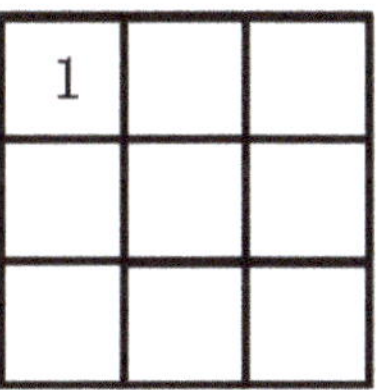

24 The weights in the figure are balanced. The same shapes have the same weight. The weight of each circular shape is 30 ounces. What is the weight in ounces of the square shape indicated by the question mark?

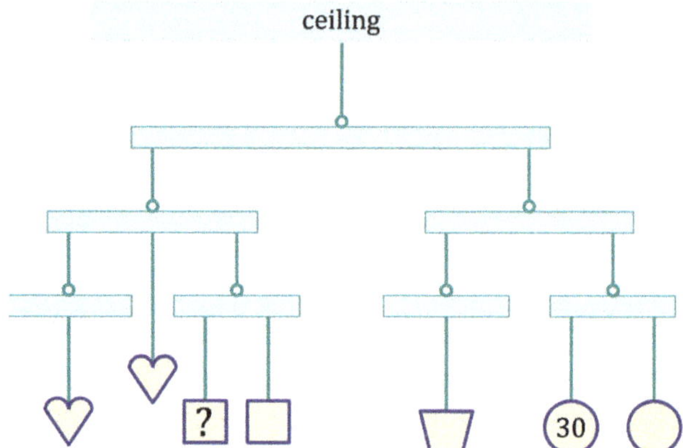

(A) 10
(B) 20
(C) 30
(D) 40
(E) 5

2008

3 Points Each

1 Ann eats 3 pieces of candy each day. How many pieces of candy does she eat in a week?

(A) 7
(B) 18
(C) 21
(D) 28
(E) 37

---

2 An adult ticket to the zoo costs $4, and a ticket for a child is $1 cheaper. On a certain Sunday, a father went to the zoo with his two children. How much did they have to pay for the tickets?

(A) $5
(B) $6
(C) $7
(D) $10
(E) $12

---

3 Luke gave the bouquets of flowers shown below to his mother, grandmother, aunt, and two sisters. Which of the bouquets did his mother receive, if we know that the flowers his aunt and sisters received were the same color, and that his grandmother did not receive roses?

(A)  yellow tulips

(B)  pink roses

(C) 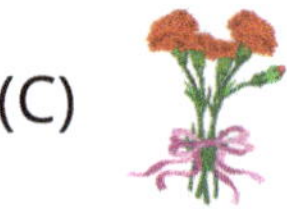 red carnations

(D)  yellow roses

(E) 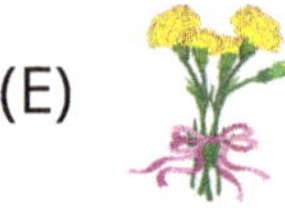 yellow carnations

4 Adelaide has 37 CDs. Her friend Mary said, "If you give me 10 of your CDs, then we will have the same number of CDs." How many CDs did Mary have?

(A) 10
(B) 17
(C) 22
(D) 27
(E) 32

---

5 Jack drew a point on a piece of paper. Next, he drew four different straight lines going through this point. Into how many pieces did these lines divide the paper?

(A) 4
(B) 6
(C) 5
(D) 8
(E) 12

---

6 In 6 hours and 30 minutes the clock will show 4 o'clock. What time does the clock show now?

(A) 9:30
(B) 4:00
(C) 8:00
(D) 2:30
(E) 10:30

7 Charlie is playing with two identical cards which are equilateral triangles, as shown. He places them on a clean piece of paper, either partly on top of each other or touching each other, and then he traces the figure. Which figure can he not get in this way?

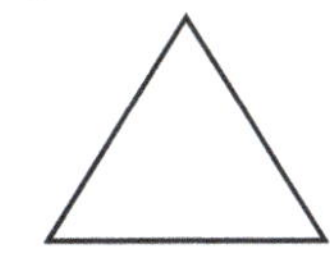 

(A) 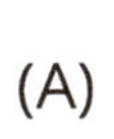 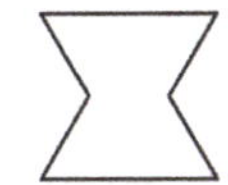

(B) 

(C) 

(D) 

(E) 

---

8 The storm made a hole in one side of the roof (see the picture). There were 10 roof tiles in each of the 7 rows. How many tiles are left on this side of the roof?

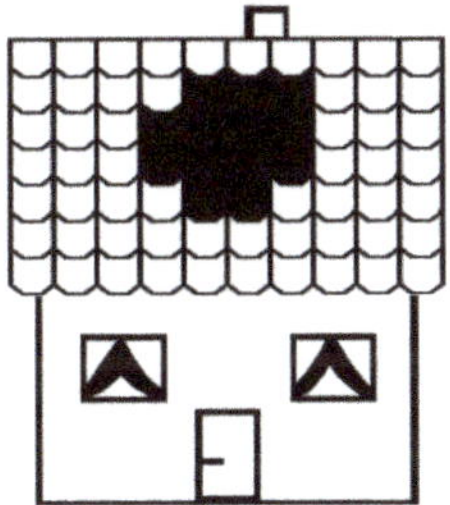

(A) 57
(B) 59
(C) 61
(D) 67
(E) 70

4 Points Each

9 In 2008, the Math Kangaroo competition is taking place in some school for the seventeenth time. Maggie took part in the seventh Math Kangaroo when she was 10 years old. In what year was Maggie born?

(A) 1986
(B) 1987
(C) 1998
(D) 1990
(E) 1988

---

10 Greg likes to multiply by 3, Jim likes to add 2, and Michael likes to subtract 1. In what order should the boys perform their favorite operations, each boy only once, so that starting with the number 3 they end up with 14?

(A) Greg, Jim, Michael
(B) Michael, Greg, Jim
(C) Greg, Michael, Jim
(D) Jim, Greg, Michael
(E) Michael, Jim, Greg

---

11 Grace is taller than Ann, but shorter than Tania. Irena is taller than Kate, but shorter than Grace. Which of the girls is the tallest?

(A) Grace
(B) Ann
(C) Kate
(D) Irena
(E) Tania

12 Each of the figures A to E shown below is made out of 5 blocks. Which of the figures can you not get from the figure on the right if you move exactly one cube?

(A) 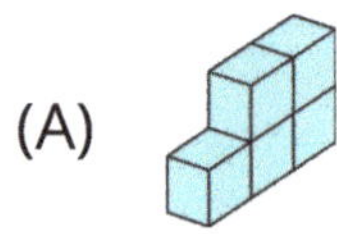

(B) 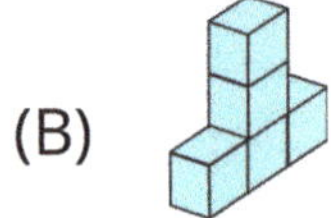

(C) 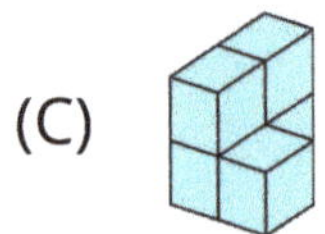

(D) 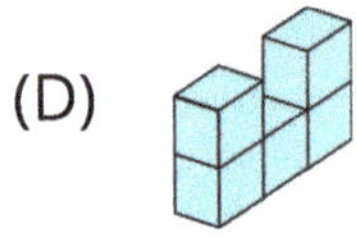

(E) 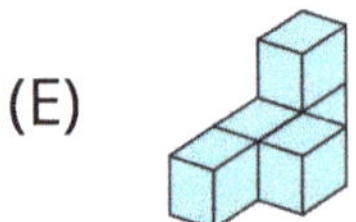

---

13 For every two different numbers with the sum equal to 45, at least one of them is less than:

(A) 5
(B) 18
(C) 20
(D) 22
(E) 23

---

14 A certain small hotel can house 21 guests. There are 5 three-person rooms and a certain number of two-person rooms. How many two-person rooms are there?

(A) 1
(B) 2
(C) 3
(D) 4
(E) 6

15 There are three songs on a certain CD. The first song is 6 minutes and 25 seconds long, the second song is 12 minutes and 25 seconds long, and the third song is 10 minutes and 13 seconds long. How long does it take to play the whole CD?

(A) 28 minutes 30 seconds
(B) 29 minutes 3 seconds
(C) 30 minutes 10 seconds
(D) 31 minutes 13 seconds
(E) 31 minutes 23 seconds

---

16 Lynn shot 2 arrows at the target and got 5 points—see the picture. How many different scores can one get if both arrows hit the target?

(A) 4
(B) 6
(C) 8
(D) 9
(E) 10

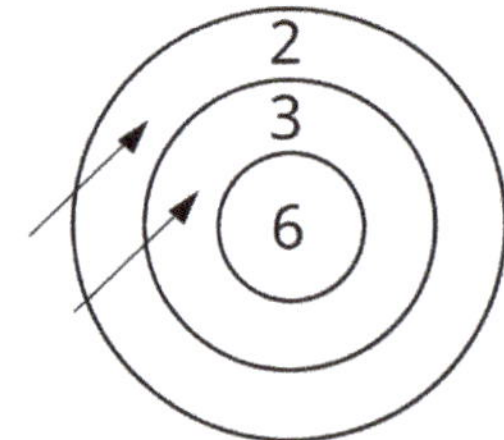

5 Points Each

17 Blocks with all right angles with the dimensions 1 cm × 2 cm × 4 cm are packed into cubes with the dimensions 4 cm × 4 cm × 4 cm. How many blocks make one of these cubes?

(A) 6
(B) 7
(C) 8
(D) 9
(E) 10

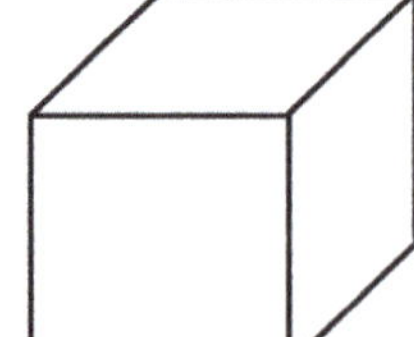

18 A kangaroo noticed that each winter he gains 5 kilograms of weight, and each summer he loses 4 kilograms. During the spring and fall, his weight does not change. In the spring of 2008, he weighed 100 kilograms. How much did he weigh in the fall of 2004?

(A) 92 kilograms
(B) 93 kilograms
(C) 94 kilograms
(D) 96 kilograms
(E) 98 kilograms

---

19 A square garden was divided into four parts: pool (P), flowerbed (F), lawn (L), and sandbox (S)—see the picture. The lawn and flowerbed are squares. The perimeter of the lawn is 20 m, and the perimeter of the flowerbed is 12 m. What is the perimeter of the pool?

(A) 10 m
(B) 12 m
(C) 14 m
(D) 16 m
(E) 18 m

P	F
L	S

---

20 A boy named Henry has as many brothers as sisters. His sister Diane has twice as many brothers as sisters. How many children are there in the family?

(A) 3
(B) 4
(C) 5
(D) 6
(E) 7

21 How many two-digit numbers are there where the ones digit is greater than the tens digit?

(A) 26
(B) 18
(C) 9
(D) 30
(E) 36

---

22 Right now, Mary is five times as old as her sister Li. In 6 years, she will be twice as old as Li. How old will Mary be in 10 years?

(A) 15
(B) 20
(C) 25
(D) 30
(E) 35

---

23 In the botanical garden shown in the picture, visitors walk only on the marked paths. In how many different ways can one go from greenhouse A to greenhouse B if you only walk on a given path once?

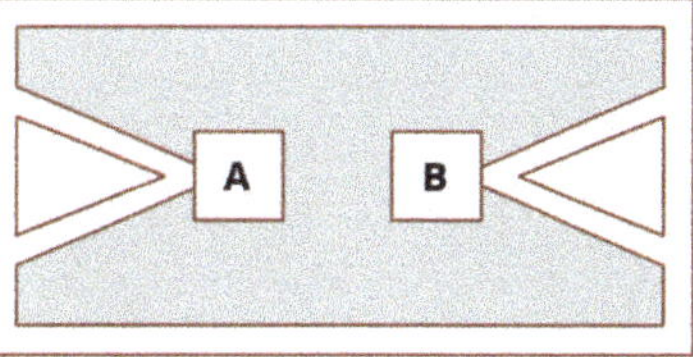

(A) 4
(B) 6
(C) 8
(D) 10
(E) 12

24 Altogether, there are 40 liters of water in two containers. First, 5 liters were poured from the first container to the second, and then enough water was poured from the second container to the first to double the amount of water in the first container. It turned out that at that point both containers ended up with the same amount of water. How much water was in the second container at the beginning?

(A) 20
(B) 35
(C) 15
(D) 25
(E) 10

2010

3 Points Each

1 Which of the numbers below is the greatest?

(A) 2 + 0 − 1 + 0
(B) 2 − 0 − 1 + 0
(C) 2 + 0 − 1 − 0
(D) 2 − 0 + 1 + 0
(E) 2 − 0 − 1 − 0

---

2 A dance lesson lasts 40 minutes, and it started at 11:50. Stan was late and came to class exactly half way through. What time did Stan come to class?

(A) 11:30
(B) 12:00
(C) 12:10
(D) 12:20
(E) 12:30

---

3 Several kids were measuring the length of a sandbox by their steps. Anna needed 15 whole steps to walk its length, Beata needed 17 whole steps, Daniel needed 12 whole steps, and Igor needed 14 whole steps. Whose steps were the longest?

(A) Anna's
(B) Beata's
(C) Daniel's
(D) Igor's
(E) It cannot be determined.

---

4 Adam spent five days preparing for a test. The first day he solved one problem, and on each consecutive day he solved twice as many problems as the day before. How many problems did Adam solve altogether preparing for the test?

(A) 15
(B) 16
(C) 31
(D) 33
(E) 63

---

5 John has 4 pieces of cardboard with this design 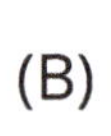. Which of the patterns below can he not make using these pieces?

(A) 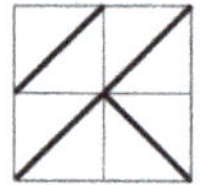

(B) 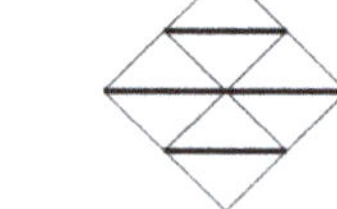

(C) 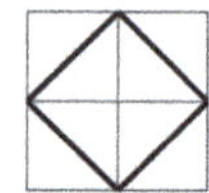

(D) 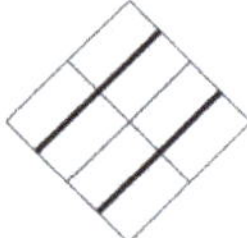

(E) 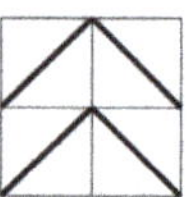

---

6 At the cosmetics store, you can buy a bar of soap for $4, a bottle of shampoo for $9, and jar of face cream for $5. You can also buy these three items as a set for $15. How much money would mother save if she bought the set instead of buying the three items separately?

(A) $3
(B) $4
(C) $5
(D) $6
(E) $7

7 Eva the centipede has 50 pairs of feet. She had shoes on some of her feet, but on the rest of her feet she did not have shoes. Today she bought 16 pairs of new shoes and put them on the feet that were without shoes. She still has 7 pairs of feet without shoes. On how many feet did she have shoes before she bought the 16 pairs of shoes?

(A) 27
(B) 40
(C) 54
(D) 70
(E) 77

---

8 The figure shown in Picture 1 was made out of six identical coins. What is the smallest number of coins that we need to move to make the figure shown in Picture 2?

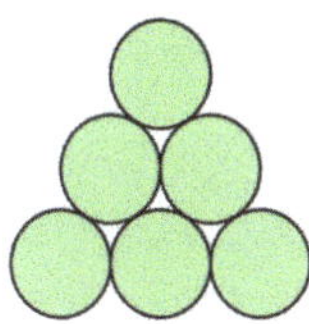
Picture 1

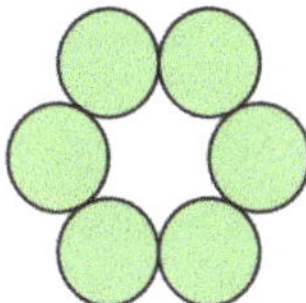
Picture 2

(A) 1
(B) 2
(C) 3
(D) 4
(E) 5

4 Points Each

9 The product $60 \times 60 \times 24 \times 7$ is equal to

(A) the number of minutes in seven weeks.
(B) the number of seconds in seven hours.
(C) the number of minutes in twenty-four weeks.
(D) the number of hours in sixty days.
(E) the number of seconds in one week.

10 Adam, Luke, Tom, and Alex went out for ice cream. Adam ate more ice cream than Alex, and Tom ate more ice cream than Luke but less than Alex. Which of the lists below gives the names of the boys in order from the one who ate the most ice cream to the one who ate the least?

(A) Adam, Tom, Luke, Alex
(B) Adam, Alex, Tom, Luke
(C) Tom, Adam, Luke, Alex
(D) Alex, Adam, Luke, Tom
(E) Tom, Luke, Adam, Alex

---

11 Matthew and Clara live in a skyscraper. Clara lives 12 floors above Matthew. One day Matthew went to visit Clara, and he took the stairs up from his apartment to Clara's apartment. Half-way up he was on the 8th floor. On what floor does Clara live?

(A) 12
(B) 14
(C) 16
(D) 20
(E) 24

---

12 We made a big cube using 64 small white cubes, and then we painted five of the sides of the big cube. How many of the small cubes have exactly two sides painted?

(A) 4
(B) 8
(C) 16
(D) 20
(E) 24

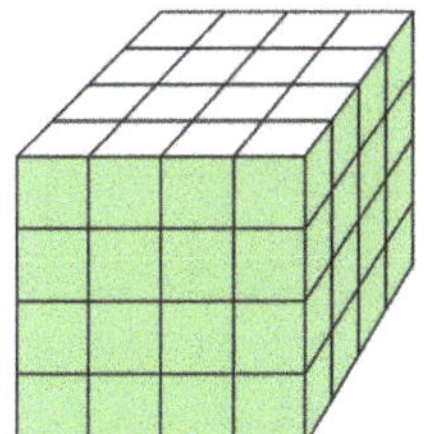

13 Grandpa has a 10-meter by 20-meter yard. One of the shorter sides of the yard is along the house. Grandpa decided to put a fence around the remaining three sides of the yard and to paint the fence. Painting 5 meters of the fence uses one-half of a two-gallon can of paint. A can of paint costs $40. How much will Grandpa pay for the paint he needs to paint the whole fence?

(A) $120
(B) $400
(C) $240
(D) $200
(E) $480

---

14 Camilla wrote all the positive integers from 1 to 100 in order on a chart with 5 columns. A part of the chart is shown in the picture. Her brother cut out a part of the chart and then he erased some of the numbers from it. Which picture represents the part of the incomplete chart cut out by Camilla's brother?

1	2	3	4	5
6	7	8	9	10
11	12	13	14	15
16	17	18	19	20

(A)

	43			
		48		

(B)

				60
	52			

(C)

			69	
	72			

(D)

	81			
	86			

(E)

		87		
			94	

15 A square piece of paper is white on one side and green on the other side. Anne divided it into 9 little squares. She labeled some edges with natural numbers 1 to 8 (see Picture 1). What is the sum of the numbers along the edges which she cut (see Picture 2)?

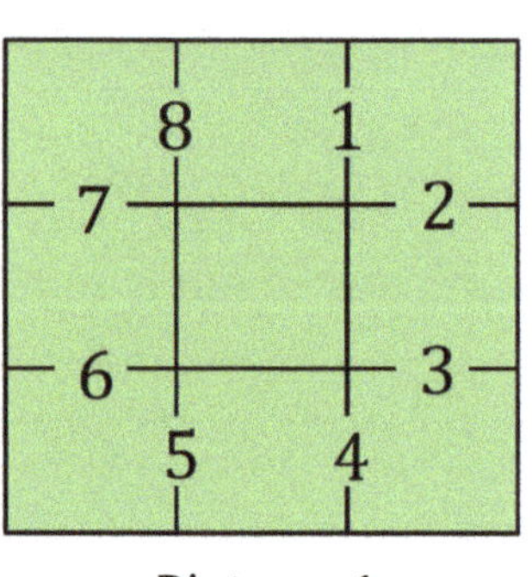

Picture 1

Picture 2

(A) 16
(B) 17
(C) 18
(D) 20
(E) 21

---

16 If the numbers in each of the two rows have the same sum, what is the value of *?

1	2	3	4	5	6	7	8	9	10	199
11	12	13	14	15	16	17	18	19	20	*

(A) 99
(B) 100
(C) 209
(D) 289
(E) 299

5 Points Each

17 Ella folded a square piece of paper twice, and so made a square with each of its sides as long as half of the original piece of paper. She then cut off all four corners from the square she made. Which of the pictures below shows the piece of paper after unfolding?

(A)

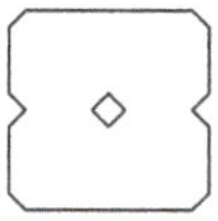

(B)

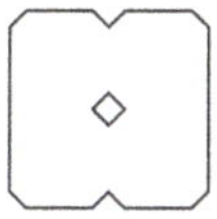

(C)

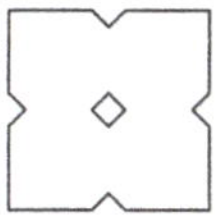

(D)

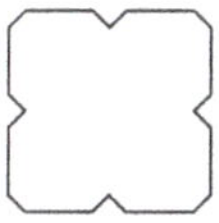

(E)

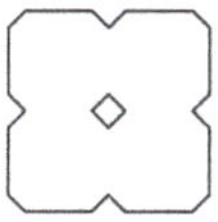

18 Anna, Beata, and Jack go to the same school. One day the librarian said to them, "Guess how many books we have in the school library." Anna said 2010, Beata said 1998, and Jack said 2015. It turned out that the number of books in the library differed from the numbers given by the children by 12, 7, and 5 (these numbers are not necessarily in the order they made their guesses). How many books are there in the library?

(A) 2005
(B) 2008
(C) 2003
(D) 2020
(E) 2022

19 Adam and Tom are walking in the same direction around a circular table and counting chairs. They begin their count with different chairs. Tom's twelfth chair is Adam's third chair, while Tom's fifth chair is Adam's eighteenth chair. How many chairs are there at the table?

(A) 20
(B) 18
(C) 30
(D) 22
(E) 23

20 A jeweler can make chains of any length using identical links. Picture 1 shows such a chain made out of three links. A single link is shown in Picture 2. What is the length of a chain made out of five links?

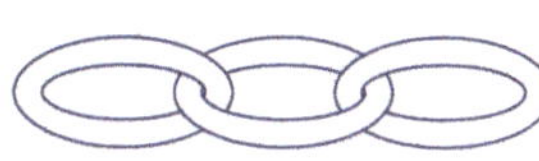

Picture 1

0.5 mm

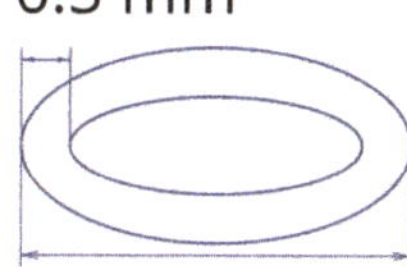

4 mm

Picture 2

(A) 20 mm
(B) 19 mm
(C) 17.5 mm
(D) 16 mm
(E) 15 mm

21 The ladybugs which live on an enchanted meadow are either red and have 6 spots, or yellow and have 10 spots. Various animals, including both red and yellow ladybugs, came to a birthday party for the dragonfly. The dragonfly noticed that the total number of spots on the ladybugs that were at the party was 42. How many ladybugs came to the dragonfly's birthday party?

(A) 10
(B) 7
(C) 6
(D) 8
(E) 5

22 Each of Basil's friends added the number of the day and the number of the month of his birthday and obtained 35. Their birthdays all fall on different days. What is the greatest possible number of friends that Basil has?

(A) 7
(B) 8
(C) 9
(D) 10
(E) 12

23 Paul, Darius, Micah, and Jeff met at a concert in Chicago, but they came from different cities: Pittsburgh, Dallas, New York, and Washington. We know that:

- Paul and the boy from Washington met in Chicago early in the morning on the day of the concert. They had never been to Pittsburgh or to New York.
- Micah is not from Washington, and he came to Chicago later than the boy from Pittsburgh.
- Jeff liked the concert more than the boy from Pittsburgh did.

What city is Jeff from?

(A) Pittsburgh
(B) New York
(C) Dallas
(D) Washington
(E) Chicago

24 Olivia is ten years old and is six times younger than her grandmother. Olivia's grandmother is 14 years older than the ages of Olivia and of her mother added together. Olivia's great grandmother's age is equal to the sum of Olivia's grandmother's and Olivia's mother's ages. How old is Olivia's great-grandmother?

(A) 106
(B) 69
(C) 70
(D) 89
(E) 96

2012

3 Points Each

1 Basil wants to write the word MATHEMATICS on a sheet of paper. He wants to color different letters with different colors, and the same letters with the same color. How many colors will he need?

(A) 7
(B) 8
(C) 9
(D) 10
(E) 13

2 In four of the five pictures below the white area is equal to the gray area. In which picture are the white area and the gray area different?

(A) 

(B) 

(C) 

(D) 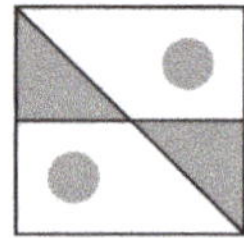

(E) 

3 Father hangs the laundry outside on a clothesline. He wants to use as few pins as possible. For 3 towels he needs 4 pins, as shown. How many pins does he need for 9 towels?

(A) 8
(B) 10
(C) 12
(D) 14
(E) 16

4 Iljo colors the squares A2, B1, B2, B3, B4, C3, D3, and D4. Which pattern does he get?

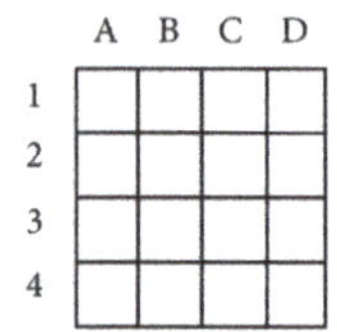

(A) 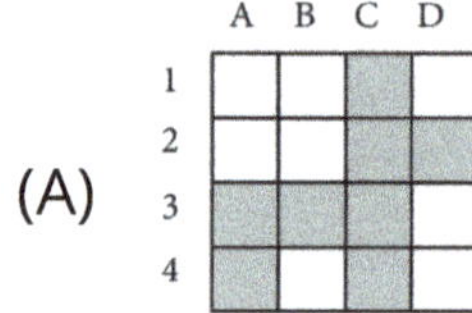

(B) 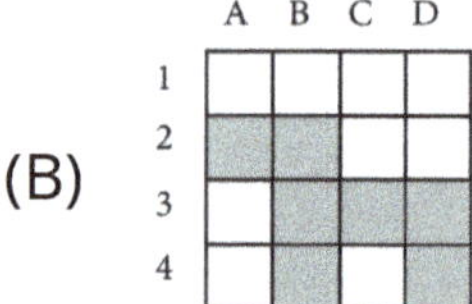

(C) 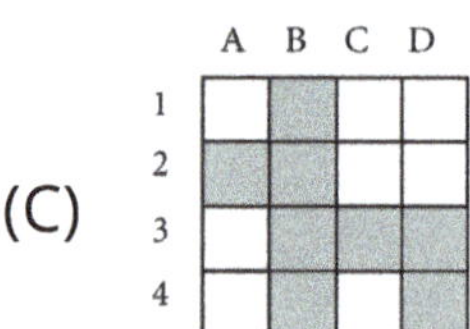

(D) 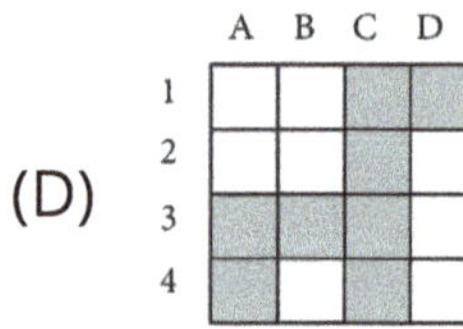

(E) 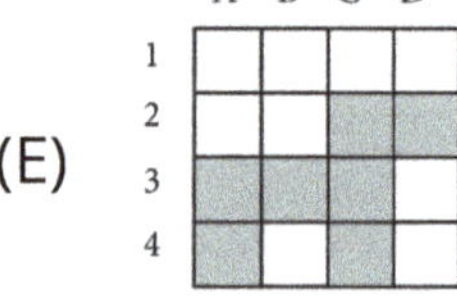

5 13 children are playing hide and seek. One of them is the "seeker" and the others hide. After a while, 9 children have been found. How many children are still hiding?

(A) 3
(B) 4
(C) 5
(D) 9
(E) 22

---

6 Mike and Jake were playing darts. Each one threw three darts (see the picture). Who won and how many points more than his opponent did he score?

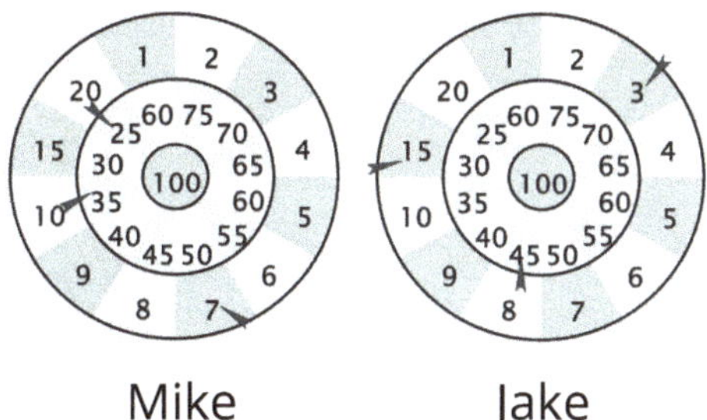

(A) Mike; he scored 3 points more.
(B) Jake; he scored 4 points more.
(C) Mike; he scored 2 points more.
(D) Jake; he scored 2 points more.
(E) Mike; he scored 4 points more.

---

7 A regular rectangular pattern on a wall was created with 2 kinds of tiles: gray and striped. Some tiles have fallen off the wall (see the picture). How many gray tiles have fallen off?

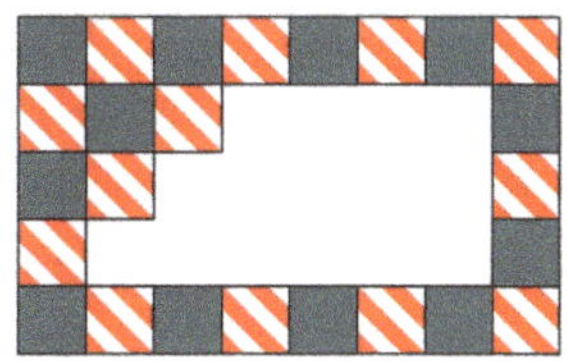

(A) 9
(B) 8
(C) 7
(D) 6
(E) 5

---

8 The year 2012 is a leap year, which means that there are 29 days in February. Today, on March 15, 2012, my grandfather's ducklings are 20 days old. When did they hatch from their eggs?

(A) on February 19
(B) on February 21
(C) on February 23
(D) on February 24
(E) on February 26

4 Points Each

9 You have L-shaped tiles, each consisting of 4 squares as shown:

How many of the following shapes can you make by gluing together two of these tiles?

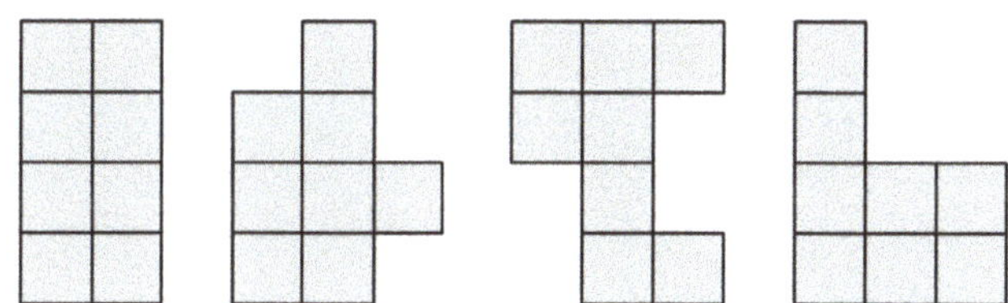

(A) 0
(B) 1
(C) 2
(D) 3
(E) 4

---

10 Three balloons cost 12 cents more than one balloon. How many cents does one balloon cost?

(A) 4
(B) 6
(C) 8
(D) 10
(E) 12

11 Grandmother made 20 gingerbread cookies for her grandchildren. She decorated them with raisins and nuts. First, she decorated 15 cookies with raisins and then she decorated 15 cookies with nuts. At least how many cookies were decorated with both raisins and nuts?

(A) 4
(B) 5
(C) 6
(D) 8
(E) 10

---

12 In a Sudoku puzzle the numbers 1, 2, 3, 4 can occur only once in each column and in each row. In the mathematical sudoku to the left, Patrick first writes in the results of the calculations. Then he completes the sudoku. Which number will Patrick put in the gray cell?

1×1		1×3	
2×2	$6 - 3$		$6 - 5$
$4 - 1$	$1 + 3$	$8 - 7$	
$9 - 7$	$2 - 1$		(gray)

(A) 1
(B) 2
(C) 3
(D) 4
(E) 1 or 2

---

13 Among Nikolay's classmates there are twice as many girls as boys. Which of the following numbers can be equal to the number of all the children in this class?

(A) 30
(B) 20
(C) 24
(D) 25
(E) 29

---

14 In the animal school, 3 kittens, 4 ducklings, 2 baby geese, and several lambs are taking lessons. The teacher owl found that all of her pupils have 44 legs altogether. How many lambs are there among them?

(A) 6
(B) 5
(C) 4
(D) 3
(E) 2

---

15 A rectangular prism is made of four pieces, as shown. Each piece consists of four cubes and is a single color. What is the shape of the white piece?

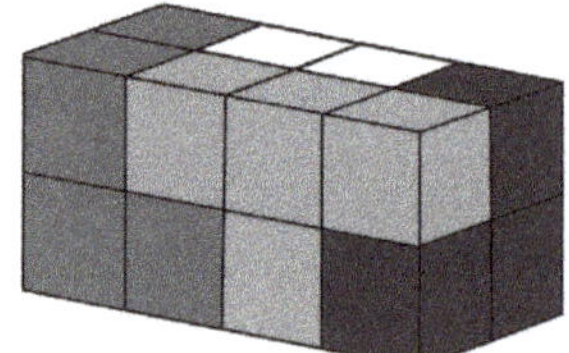

(A) 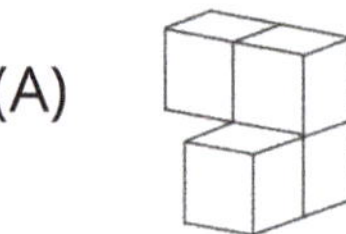

(B) 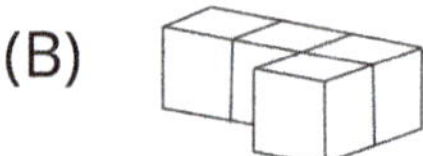

(C) 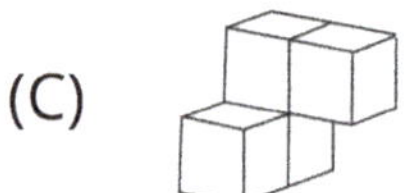

(D) 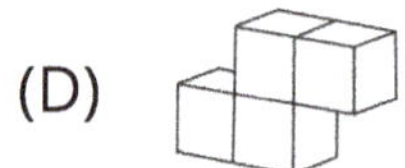

(E) 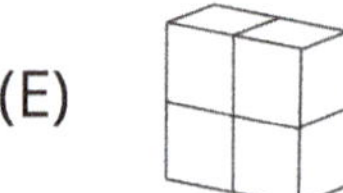

---

16 At a Christmas party there was exactly one candlestick on each of the 15 tables. There were 6 five-branched candlesticks; the rest of them were three-branched candlesticks. How many candles had to be bought for all the candlesticks?

(A) 45
(B) 50
(C) 57
(D) 60
(E) 75

5 Points Each

17 A grasshopper wants to climb a staircase with many steps. She makes only two kinds of jumps: 3 steps up or 4 steps down. Beginning at the ground level, at least how many jumps will she have to make in order to take a rest on the 22nd step?

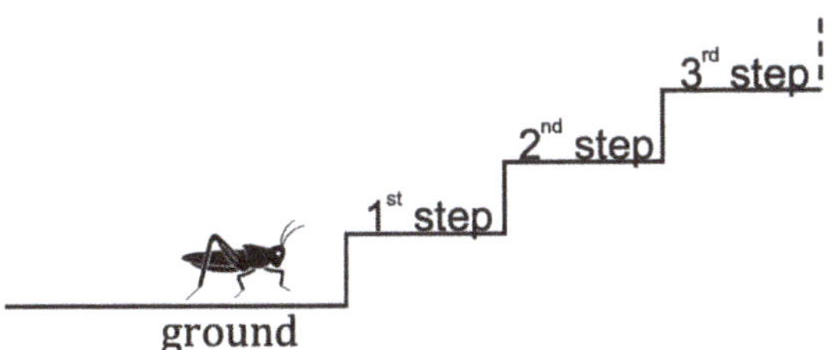

(A) 7
(B) 9
(C) 10
(D) 12
(E) 15

---

18 Frank made a domino snake out of seven tiles. He put the tiles next to each other so that the sides with the same number of dots were touching. Originally the snake had 33 dots on its back. However, his brother George took away two tiles from the snake (see the picture). How many dots were in the place with the question mark?

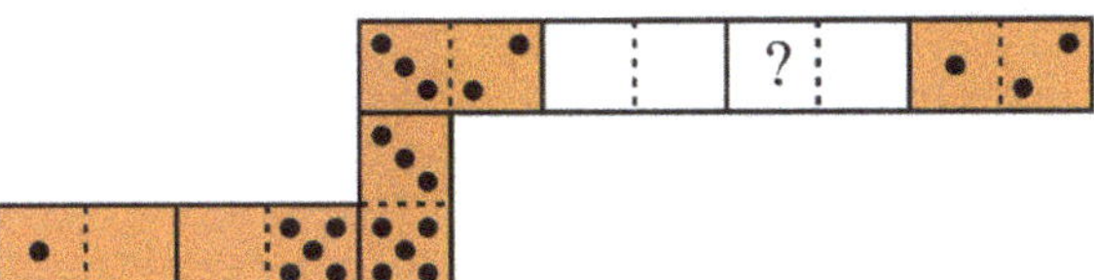

(A) 2
(B) 3
(C) 4
(D) 5
(E) 6

19 Gregory forms two numbers with the digits 1, 2, 3, 4, 5, and 6. Both numbers have three digits, and each digit is used only once. He adds these two numbers. What is the greatest sum Gregory can get?

(A) 975
(B) 999
(C) 1083
(D) 1173
(E) 1221

---

20 Laura, Iggy, Val, and Kate want to be in one photo together. Kate and Laura are best friends and they want to stand next to each other. Iggy wants to stand next to Laura because he likes her. In how many different ways can they pose for the photo if they all stand in one row?

(A) 3
(B) 4
(C) 5
(D) 6
(E) 7

21 A special clock has 3 hands of different length (one for hours, one for minutes, and one for seconds). We do not know which hand is which, but we know that the clock runs correctly. At 12:55:30 p.m. the hands were in position shown on the right. What will this clock look like at 8:11:00 p.m.?

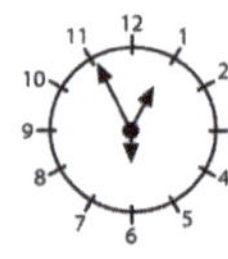

(A) 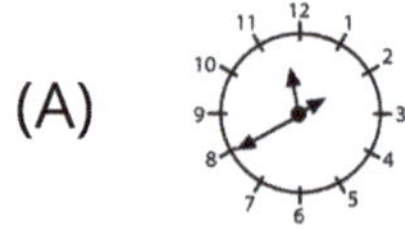

(B) 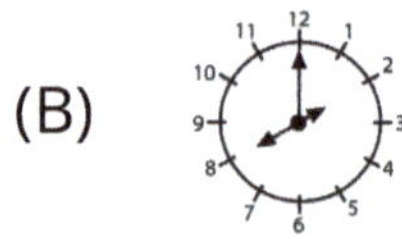

(C) 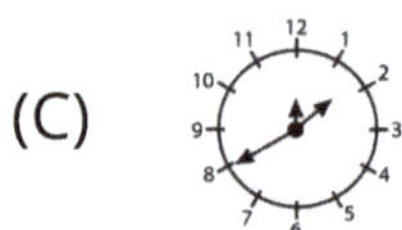

(D) 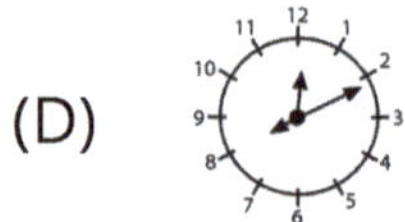

(E) 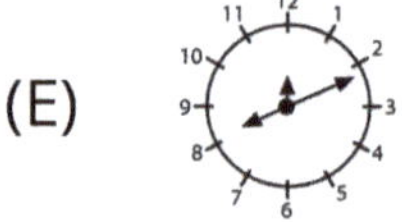

22 Michael chose a positive number, multiplied it by itself, added 1, multiplied the result by 10, added 3, and multiplied the result by 4. His final answer was 2012. What number did Michael choose?

(A) 11
(B) 9
(C) 8
(D) 7
(E) 5

23 A rectangular paper sheet measures 192 × 84 mm. You cut the sheet along just one straight line to get two parts, one of which is a square. Then you do the same with the non-square part of the sheet, and so on. What is the length of the side of the smallest square you can get in this way?

(A) 1 mm
(B) 4 mm
(C) 6 mm
(D) 10 mm
(E) 12 mm

24 In a soccer game the winner gains 3 points, while the loser gains 0 points. If the game is a tie, then the two teams gain 1 point each. A certain team played 38 games and gained 80 points. Find the greatest possible number of games that the team lost.

(A) 12
(B) 11
(C) 10
(D) 9
(E) 8

2014

2014

3 Points Each

1 Which small figure could be the central part of the larger figure with the star?

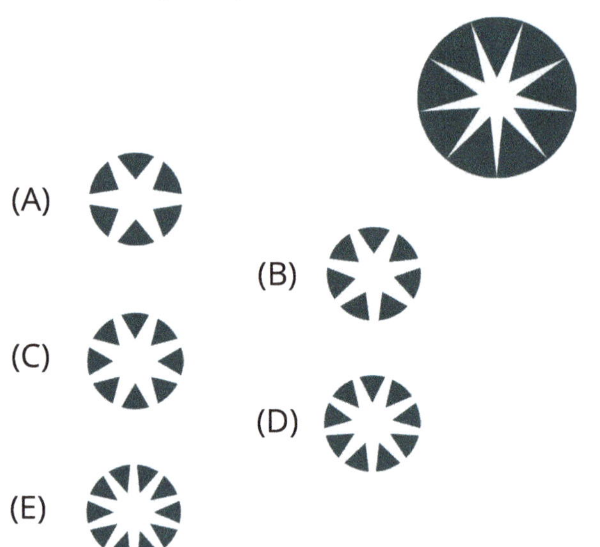

2 Jackie wants to place the digit 3 somewhere in the number 2014. Where should she place the digit 3 to make the resulting five-digit number as small as possible?

(A) in front of 2014
(B) between the 2 and the 0
(C) between the 0 and the 1
(D) between the 1 and the 4
(E) after 2014

3 Which houses are made using exactly the same triangular and rectangular pieces?

1  2 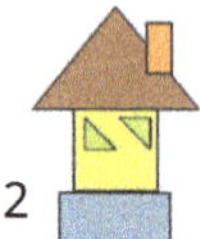 3 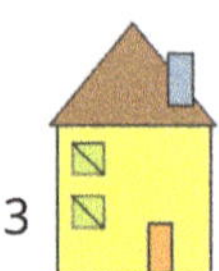 4 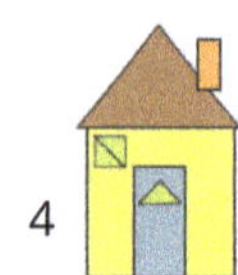 5 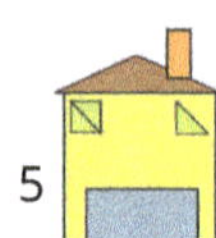

(A) 1, 4
(B) 3, 4
(C) 1, 4, 5
(D) 3, 4, 5
(E) 1, 2, 4, 5

4 When Koko the Koala is not sleeping, he eats 50 grams of leaves per hour. Yesterday he slept 20 hours. How many grams of leaves did he eat yesterday?

(A) 0
(B) 50
(C) 100
(D) 200
(E) 400

5 Maria performs the subtractions shown in the box and gets the numbers from zero to five as results. She connects the dots, starting at the dot with the result 0 and ending at the dot with the result 5, in increasing order. Which figure does she draw as a result?

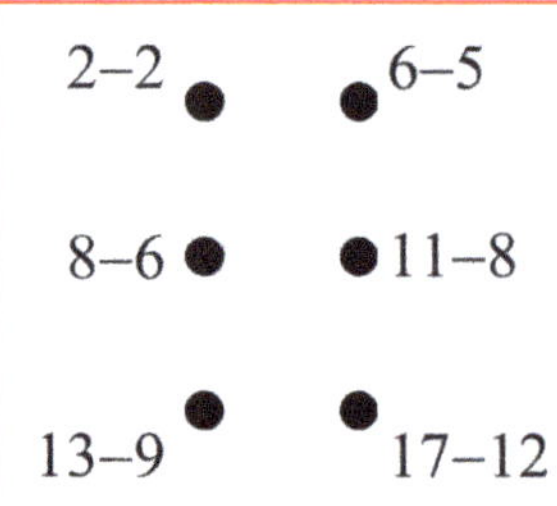

(A) 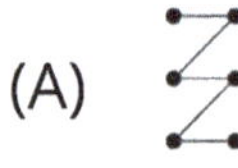

(B) 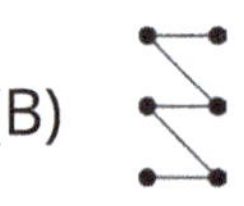

(C)

(D) 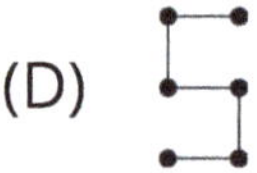

(E)

6 Adam built fewer sandcastles than Martin but more than Susan. Lucy built more sandcastles than Adam and more than Martin. Dana built more sandcastles than Martin but fewer than Lucy. Which of the children built the most sandcastles?

(A) Martin
(B) Adam
(C) Susan
(D) Dana
(E) Lucy

7 Monica writes numbers in the diagram so that each number is the product of the two numbers below it. Which number does she write in the gray cell?

(A) 0
(B) 1
(C) 2
(D) 4
(E) 8

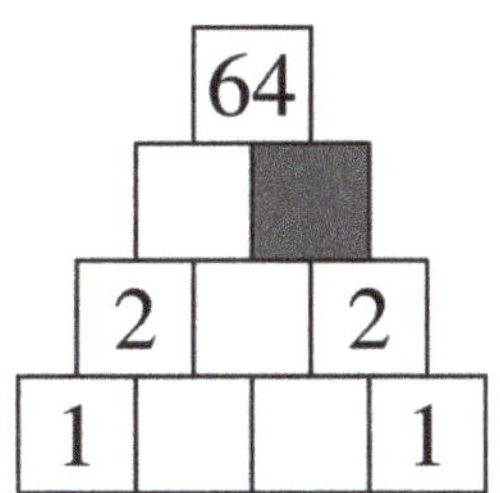

---

8 Ann has the four blue pieces shown. She needs to completely cover the white shape.

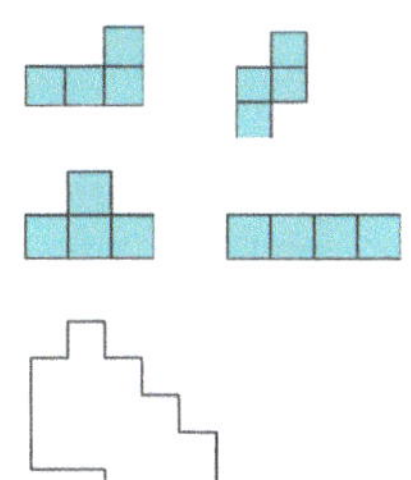

Where does she need to place the T-shaped piece so that she can use the other three pieces to cover the rest of the shape?

(A) 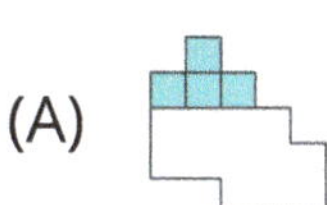

(B) 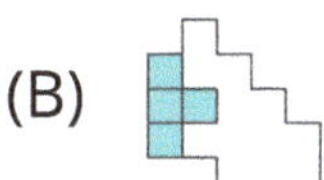

(C) 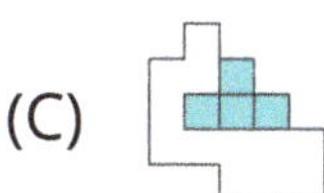

(D) 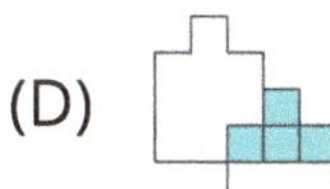

(E) 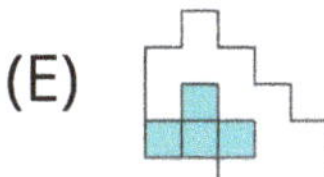

4 Points Each

9 Mr. Brown painted some flowers on the store window (see picture on the right). What do these flowers look like from the other side of the window?

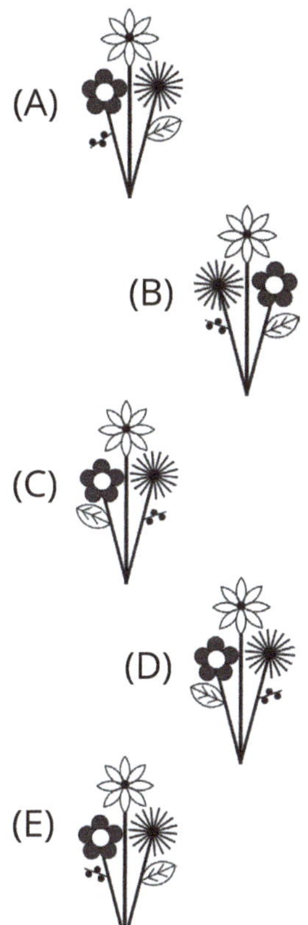

---

10 There were some pieces of candy in a bowl. Sally took half of the pieces of candy. Then Tom took half of the pieces left in the bowl. After that Clara took half of the remaining pieces. In the end there were 6 pieces of candy left in the bowl. How many pieces of candy were in the bowl at the beginning?

(A) 12
(B) 18
(C) 20
(D) 24
(E) 48

11 Which tile must be added to the picture so that the total light gray area is the same size as the total dark gray area?

(A) 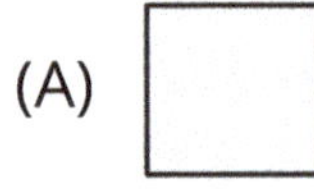

(B) 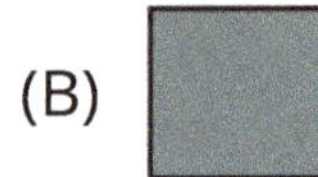

(C) 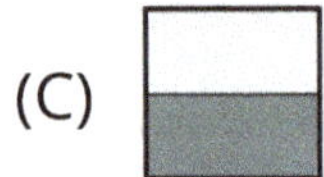

(D) 

(E) 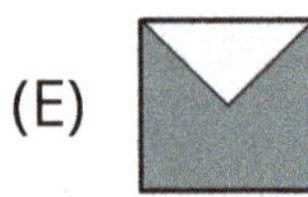

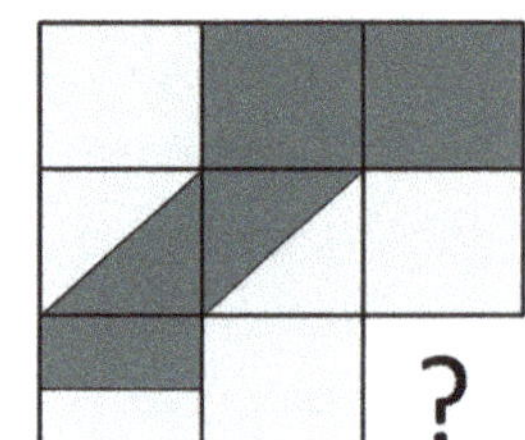

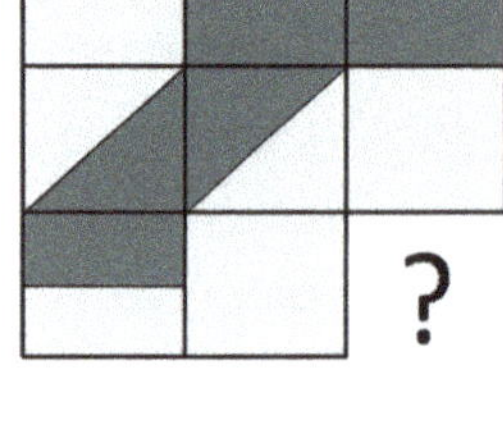

12 Paula is shooting arrows at the target shown on the right. When she misses the target, she gets zero points. Paula shoots two arrows and adds the number of points. Which of the following sums cannot be her score?

(A) 60
(B) 70
(C) 80
(D) 90
(E) 100

30 50 70

13 Mary had equal numbers of red, yellow, and green tokens. She used some of these tokens to make a pile. You can see all the tokens she used in the figure. After making the pile, she still has five tokens which were not used. How many yellow tokens did she have at the beginning?

(A) 5
(B) 6
(C) 7
(D) 15
(E) 18

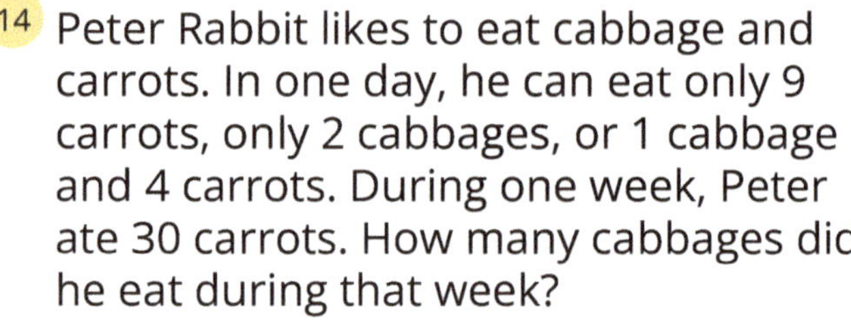

14 Peter Rabbit likes to eat cabbage and carrots. In one day, he can eat only 9 carrots, only 2 cabbages, or 1 cabbage and 4 carrots. During one week, Peter ate 30 carrots. How many cabbages did he eat during that week?

(A) 6
(B) 7
(C) 8
(D) 9
(E) 10

15 The solid in the picture was made by gluing eight identical cubes together. What does this solid look like from directly above?

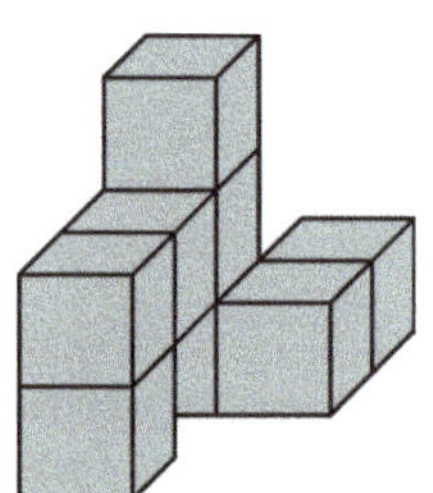

(A) 

(B) 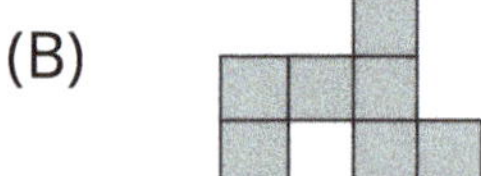

(C) 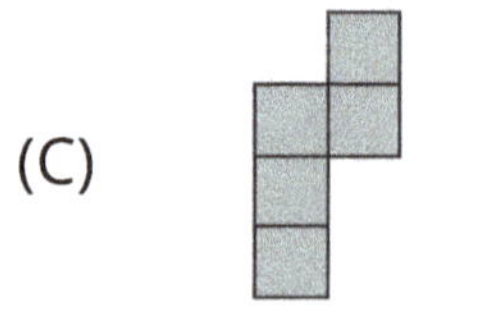

(D) 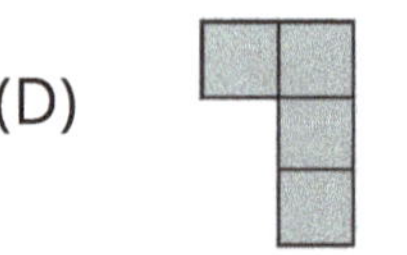

(E) 

16 How many dots are there in this picture?

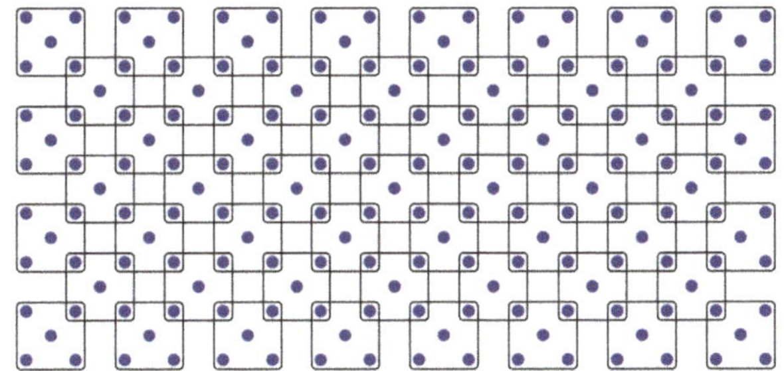

(A) 180
(B) 181
(C) 182
(D) 183
(E) 265

5 Points Each

17 On planet Kangaroo, each kangyear has 20 kangmonths, and each kangmonth has 6 kangweeks. How many kangweeks are there in one quarter of a kangyear?

(A) 9
(B) 30
(C) 60
(D) 90
(E) 120

---

18 Seven children are standing in a circle. No two boys are standing next to each other. No three girls are standing next to each other. Which of the statements below about the number of girls standing in the circle is true?

(A) 3 is the only possible number.
(B) 3 and 4 are the only possible numbers.
(C) 4 is the only possible number.
(D) 4 and 5 are the only possible numbers.
(E) 5 is the only possible number.

19 Eve arranged cards in a line as shown below. In one move, Eve can switch the places of any two cards. What is the smallest number of moves Eve needs to make to get the word KANGAROO?

(A) 2
(B) 3
(C) 4
(D) 5
(E) 6

---

20 We are making a sequence of triangles out of diamonds. The first three steps are shown. In each step a line of diamonds is added to the bottom. In the bottom line the two outside diamonds, one on each side, are white. All the other diamonds in the triangle are black. How many black diamonds will the figure have in step 6?

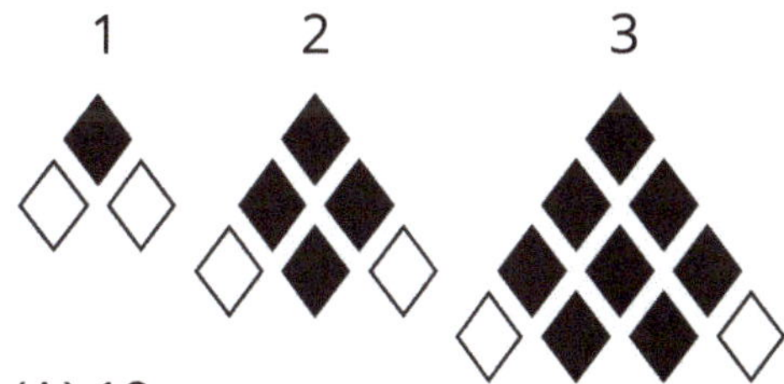

(A) 19
(B) 21
(C) 26
(D) 28
(E) 34

21 Kangaroo Hamish bought some of the toys shown in the picture and gave the cashier 150 Kangcoins. He received 20 Kangcoins back. Then he changed his mind and exchanged one of the toys for another. He got back an additional 5 Kangcoins. What toys did Hamish leave the store with?

(A) the bear and the horse
(B) the bear and the train
(C) the bear and the duck
(D) the ball and the duck
(E) the train, the ball, and the duck

---

22 Write each of the numbers 0, 1, 2, 3, 4, 5, and 6 in the squares to make the addition below correct. Which digit will be in the gray square?

(A) 2
(B) 3
(C) 4
(D) 5
(E) 6

23 What is the largest number of small squares that can be shaded in the figure below so that no square like the one shown on the right, made of four small shaded squares, appears in the figure?

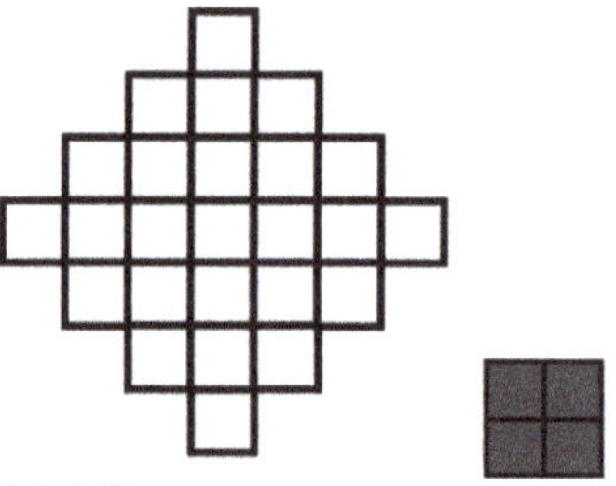

(A) 18
(B) 19
(C) 20
(D) 21
(E) 22

---

24 Nick wrote each of the numbers from 1 to 9 in the cells of the 3 × 3 table. Only four of the numbers can be seen in the figure. Nick noticed that for the number 5, the sum of the numbers in the neighboring cells is equal to 13 (neighboring cells are cells that share a side). He noticed that the same is also true for the number 6. Which number did Nick write in the shaded cell?

1		2
	(shaded)	
4		3

(A) 5
(B) 6
(C) 7
(D) 8
(E) 9

2016

3 Points Each

1 Amy, Bert, Carl, Doris, and Ernst each rolled two dice and added the number of dots.

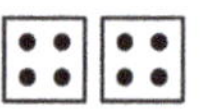

Amy Bert Carl Doris Ernst

Who rolled the largest total?

(A) Amy
(B) Bert
(C) Carl
(D) Doris
(E) Ernst

---

2 Little Kanga is 7 weeks and 2 days old. In how many days will Little Kanga be 8 weeks old?

(A) 1
(B) 2
(C) 3
(D) 4
(E) 5

---

3 Which number should be placed in the box with the question mark?

17 + 3 **20 − 16**

+

?

(A) 24
(B) 28
(C) 36
(D) 56
(E) 80

4 What does Pipo the Clown see when he looks at himself in the mirror?

(A) 

(B) 

(C) 

(D) 

(E) 

---

5 Geoff goes with his father to a circus. Their seats are numbered 71 and 72. Which way should they go?

(A) 

(B) 

(C)

(D) 

(E) 

- seats 1 to 20
- seats 21 to 40
- seats 41 to 60
- seats 61 to 80
- seats 81 to 100

6 Anna shares some apples between herself and 5 friends. Everyone gets half of an apple. How many apples does she share?

(A) 2 and a half
(B) 3
(C) 4
(D) 5
(E) 6

---

7 A rectangle is partly hidden behind a curtain. What shape is the hidden part?

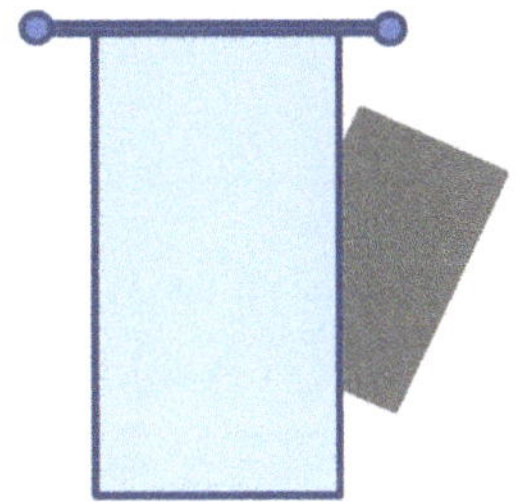

(A) a triangle
(B) a square
(C) a hexagon
(D) a circle
(E) a rectangle

---

8 Which one of the following sentences correctly describes the picture?

(A) There are as many circles as squares.
(B) There are fewer circles than triangles.
(C) There are twice as many circles as triangles.
(D) There are more squares than triangles.
(E) There are two triangles more than circles.

4 Points Each

9 The sum of the digits of the year 2016 is equal to 9. What is the next year, after 2016, where the sum of the digits of the year is equal to 9 again?

(A) 2007
(B) 2025
(C) 2034
(D) 2108
(E) 2134

---

10 The mouse wants to escape from the maze. How many different paths can the mouse take without passing through the same gate more than once?

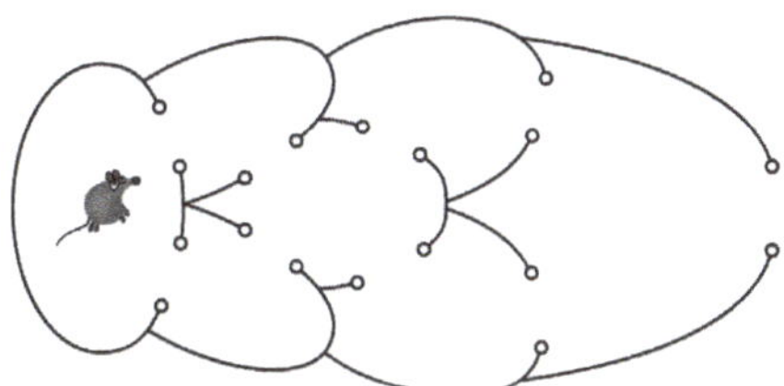

(A) 2
(B) 4
(C) 5
(D) 6
(E) 7

---

11 Zoe has two cards. She wrote a number on each side of each card. The sum of the two numbers on the first card is equal to the sum of the two numbers on the second card. The sum of the four numbers is 32. What are the two numbers on the sides that we cannot see?

(A) 7 and 0
(B) 8 and 1
(C) 11 and 4
(D) 9 and 2
(E) 6 and 3

5 **12**

12 Which tile fits in the middle in such a way that only lines with the same color touch each other?

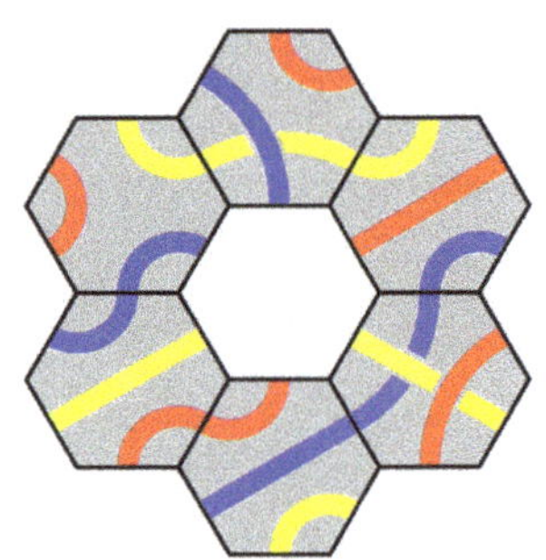

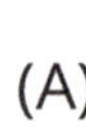 (A) 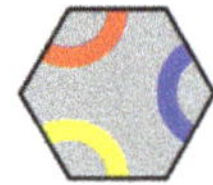

 (B) 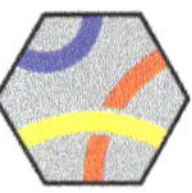

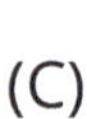 (C) 

 (D) 

 (E) 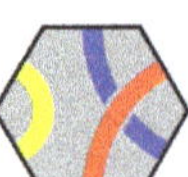

13 Each of five children had a paper square, a paper triangle, and a paper circle. Each child placed their own papers in a pile, as shown in the pictures. How many children placed the triangle above the square?

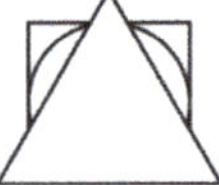  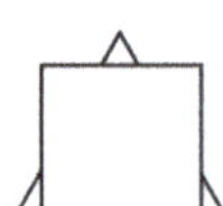 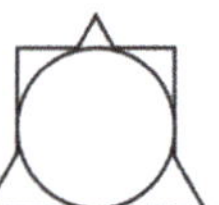 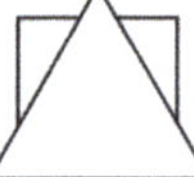

(A) 0
(B) 1
(C) 2
(D) 3
(E) 4

14 Which three of the five jigsaw pieces shown below can be joined together to form a square?

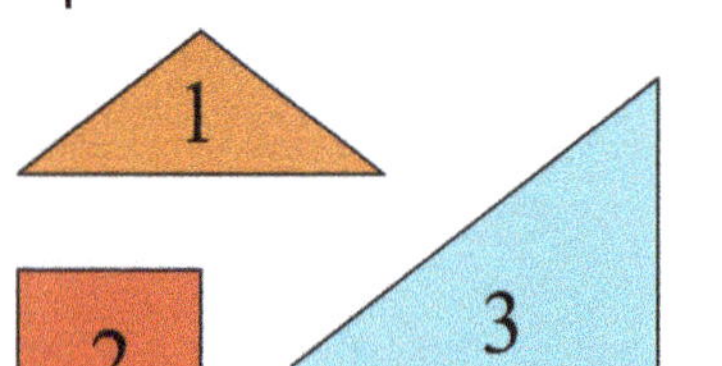

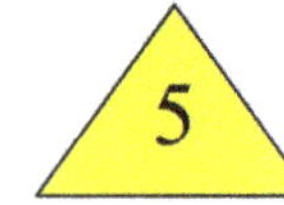

(A) 1, 3, and 5
(B) 1, 2, and 5
(C) 1, 4, and 5
(D) 3, 4, and 5
(E) 2, 3, and 5

15 Loes has started to write some numbers in the table. She decided that each row and column will contain the numbers 1, 2, and 3 exactly once. What is the sum of the numbers that she will write in the two shaded squares?

(A) 2
(B) 3
(C) 4
(D) 5
(E) 6

16 John has a board with 11 squares. He puts a coin in each of eight neighboring squares without leaving any empty squares between the coins. What is the maximum number of squares in which one can be sure that there is a coin?

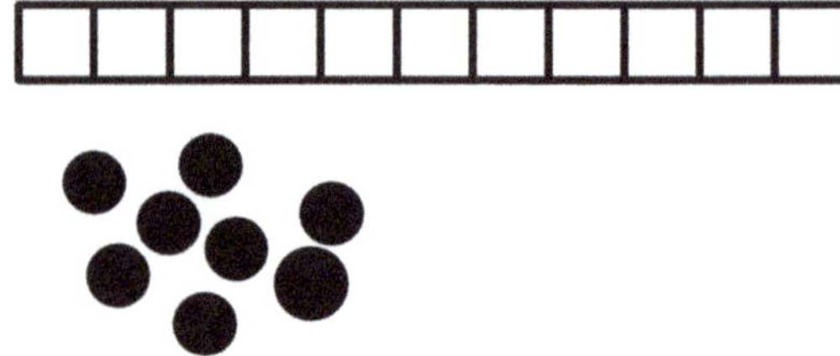

(A) 1
(B) 3
(C) 4
(D) 5
(E) 6

5 Points Each

17 When we turn over the card over its right edge, we see the result in the figure. What will we see if we turn this card over its upper edge?

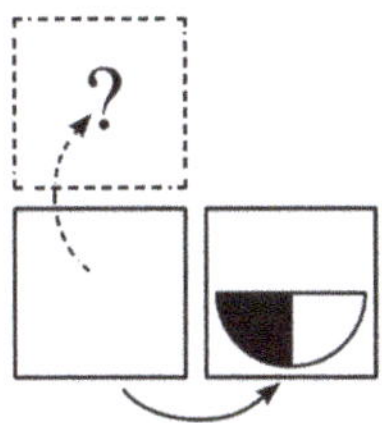

(A)

(B)

(C)

(D)

(E)

---

18 Tim, Tom, and Jim are triplets (three brothers born on the same day). Their brother Paul is exactly 3 years older. Which of the following numbers can be the sum of the ages of the four brothers?

(A) 25
(B) 27
(C) 29
(D) 30
(E) 60

---

19 Magic trees grow in a magic garden. Each tree has either 6 pears and 3 apples, or 8 pears and 4 apples. There are 25 apples in the garden. How many pears are there in the garden?

(A) 35
(B) 40
(C) 45
(D) 50
(E) 56

---

20 My dogs have 18 more legs than noses. How many dogs do I have?

(A) 4
(B) 5
(C) 6
(D) 8
(E) 9

---

21 Karin wants to place five bowls on a table in order of their weight. She already placed bowls Q, R, S, and T in order. Bowl T weighs the most.

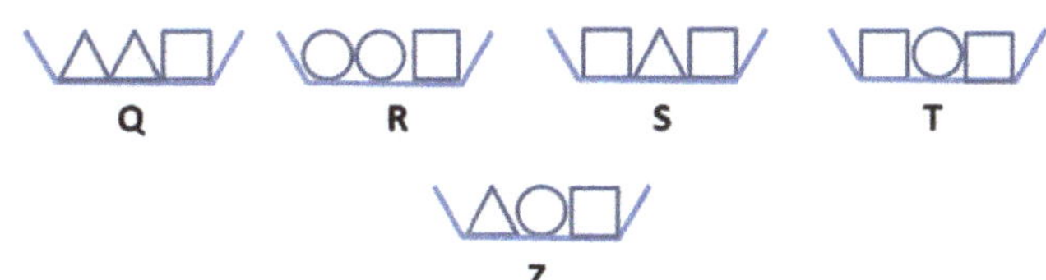

Where must she place bowl Z?

(A) to the left of bowl Q
(B) between bowl Q and bowl R
(C) between bowl R and bowl S
(D) between bowl S and bowl T
(E) to the right of bowl T

22 Rachel adds seven numbers and gets 2016. One of the numbers she adds is 201. She replaces the number 201 with 102. What sum does she get?

(A) 1815
(B) 1914
(C) 1917
(D) 2115
(E) 2118

---

23 Malte built a bar of 27 bricks. He breaks the bar into two bars in such a way that one of them is twice the length of the other. Then he takes one of the new bars and breaks it the same way. He continues in this way. Which of the following bars will he not be able to get?

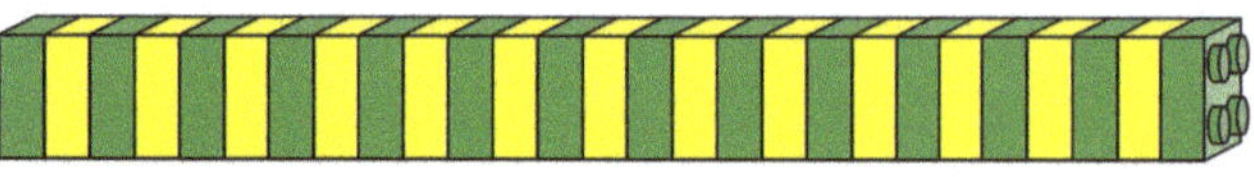

(A) 2
(B) 4
(C) 6
(D) 8
(E) 10

24 Five sparrows are sitting on a branch, as shown in the figure. Each sparrow chirps the same number of times as the number of sparrows it sees. For example, David chirps three times. Then, one sparrow turns to look in the opposite direction. Again, each of the sparrows chirps the same number of times as the number of sparrows it sees. This time, the total number of chirps is more than the first time. Which of the sparrows turned to look in the opposite direction?

(A) Angel
(B) Bertha
(C) Charlie
(D) David
(E) Eglio

2018

3 Points Each

1 Lena has 10 rubber stamps. Each stamp has one of the digits: 0, 1, 2, 3, 4, 5, 6, 7, 8 and 9. She stamps the date of the Kangaroo contest:

0 3 1 5 2 0 1 8

How many of the stamps does she use?

(A) 5
(B) 6
(C) 7
(D) 9
(E) 10

2 The picture shows 3 arrows that are flying and 9 balloons that can't move. When an arrow hits a balloon, the balloon pops, and the arrow keeps flying in the same direction. How many balloons will be hit by the flying arrows?

(A) 2
(B) 3
(C) 4
(D) 5
(E) 6

3 Susan is 6 years old. Her sister is one year younger, and her brother is one year older. What is the sum of the ages of the three siblings?

(A) 10
(B) 15
(C) 18
(D) 21
(E) 30

4 The picture shows five screws in a block. Four of the screws are the same length. One screw is shorter. Which screw is the short one?

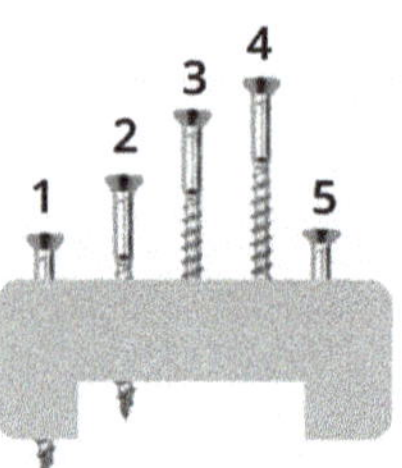

(A) 1
(B) 2
(C) 3
(D) 4
(E) 5

5 Here is Sophie the ladybug. She turns around. Which of the ladybugs below is not Sophie?

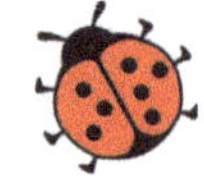

(A)
(B)
(C)
(D)
(E)

6 Lucy folds a sheet of paper in half. Then she cuts a piece out of it as shown here. What will she see when she unfolds the paper?

(A)
(B)
(C)
(D)
(E)

7 On her first turn, Diana scored 12 points total with three arrows. On her second turn she scored 15 points. How many points did she score on her third turn?

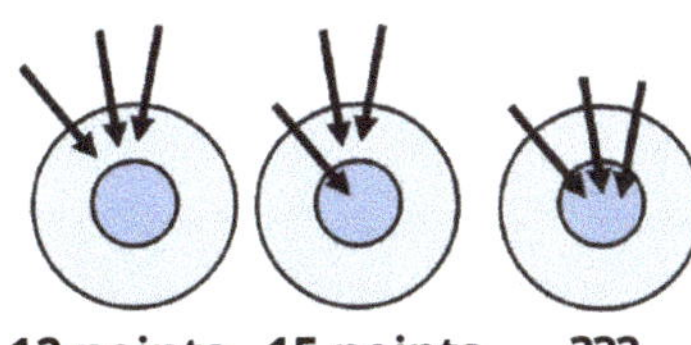

12 points **15 points** **???**

(A) 18
(B) 19
(C) 20
(D) 21
(E) 22

---

8 Mike sets the table for 8 people. He must set the table correctly for each person sitting at the table. Setting the table correctly means that the fork is on the left of the plate and the knife is on the right of the plate. For how many people did Mike set the table correctly?

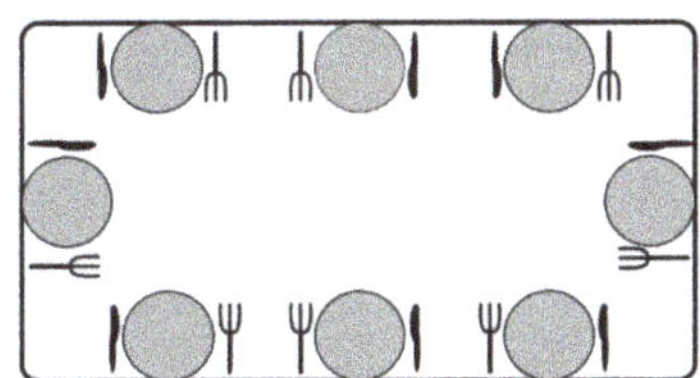

(A) 5
(B) 4
(C) 6
(D) 2
(E) 3

4 Points Each

9 Roberto makes designs using tiles like this:

How many of the 5 designs can he make?

(A) 1
(B) 2
(C) 3
(D) 4
(E) 5

---

10 Albert fills the grid with these five figures:

Each figure appears exactly once in every column and every row. Which figure must Albert put in the cell with the question mark?

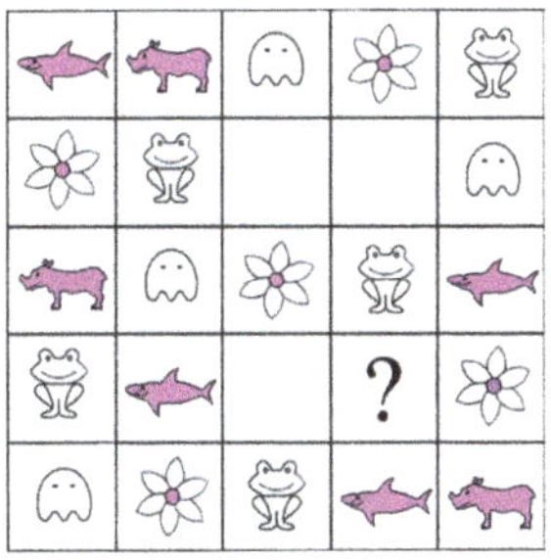

?

(A) 
(B) 
(C) 
(D) 
(E)

11 Tom cuts two kinds of shapes out of grid paper as shown. What is the smallest number of shapes that Tom needs in order to exactly cover the boat in the picture?

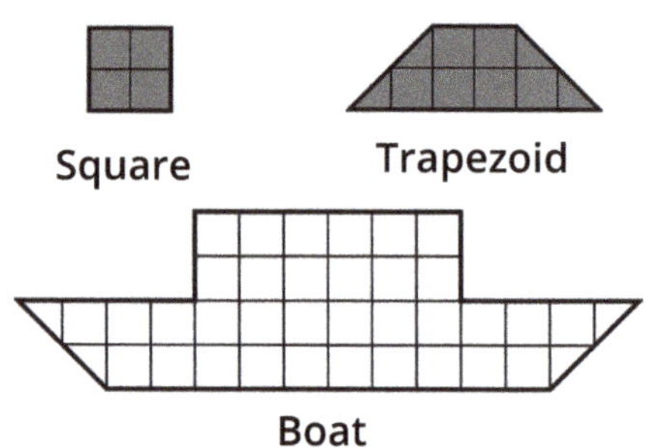

(A) 5
(B) 6
(C) 7
(D) 8
(E) 9

12 The colors in this picture are switched. Then the picture is rotated. What does the picture look like now?

(A) 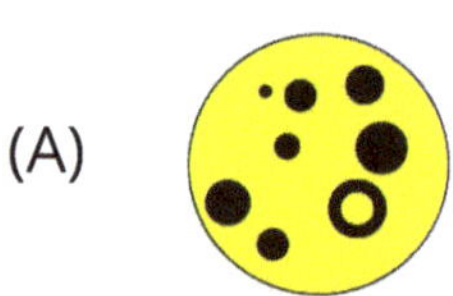

(B) 

(C) 

(D) 

(E) 

13 Peta Rabbit has 20 carrots. She eats 2 carrots every day. She ate the 12th carrot on Wednesday. On which day did she start eating the carrots?

(A) Monday
(B) Tuesday
(C) Wednesday
(D) Thursday
(E) Friday

14 Toby glues 10 cubes together to make the structure shown. He paints the whole structure, even the bottom. How many cubes are painted on exactly 4 of their faces?

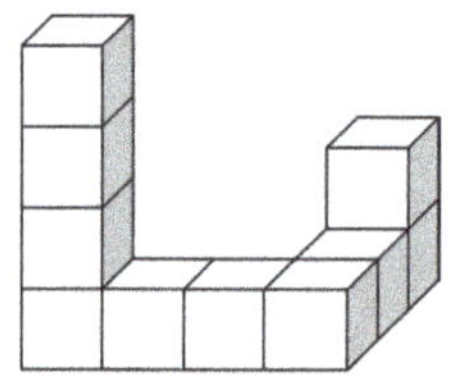

(A) 6
(B) 7
(C) 8
(D) 9
(E) 10

15 There are 8 flowers on a rose bush. Some butterflies and some dragonflies are sitting on the flowers. There is no more than one insect on each flower. More than half of the flowers have an insect on them. The number of butterflies on the flowers is twice the number of dragonflies on the flowers. How many butterflies are sitting on the flowers?

(A) 2
(B) 3
(C) 4
(D) 5
(E) 6

16 Captain Kook wants to sail from the island called Easter through every island on the map and back to Easter. The total journey is 100 kilometers (km) long. The direct distance between Desert and Lake is the same as the distance between Easter and Flower via Volcano. How far is it directly from Easter to Lake?

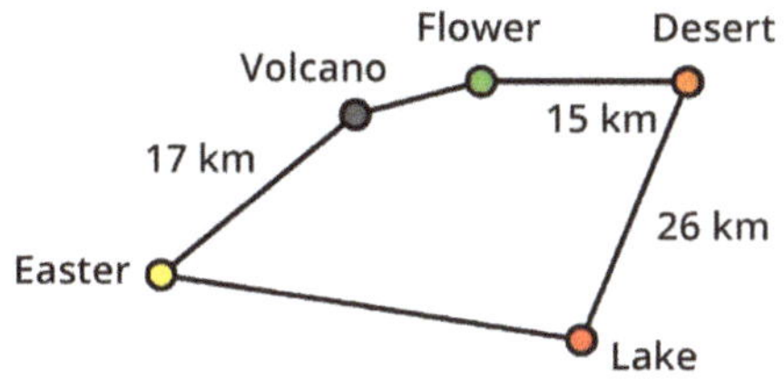

(A) 17 km
(B) 23 km
(C) 26 km
(D) 33 km
(E) 35 km

5 Points Each

17 The rooms in Kanga's house are numbered. Baby Roo enters the main door, passes through some rooms and leaves the house. The numbers on the rooms that he visits are always increasing. Through which door does he leave the house?

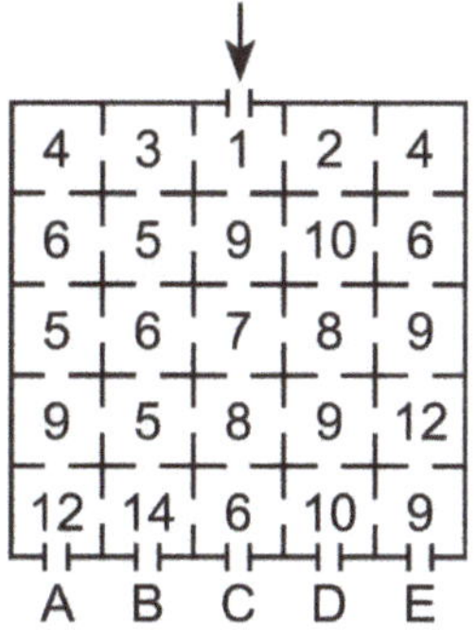

(A) A
(B) B
(C) C
(D) D
(E) E

18 Four balls each weigh 10 g, 20 g, 30 g, and 40 g. Which ball weighs 30 g?

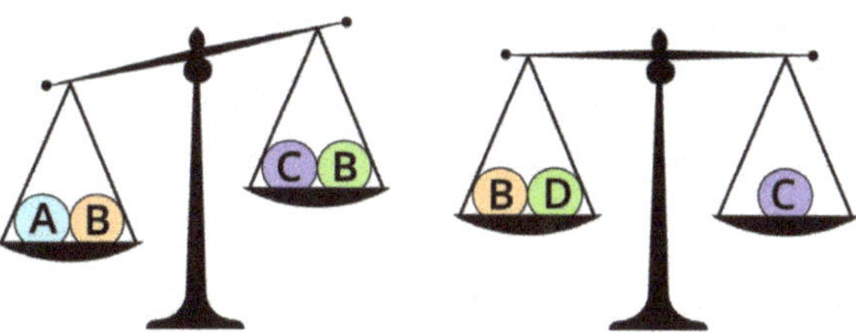

(A) A
(B) B
(C) C
(D) D
(E) It could be A or B.

19 The band shown in the drawing can be fastened in five ways. How much longer is the band fastened in one hole than the band fastened in all five holes?

Unfastened band

Band fastened in one hole

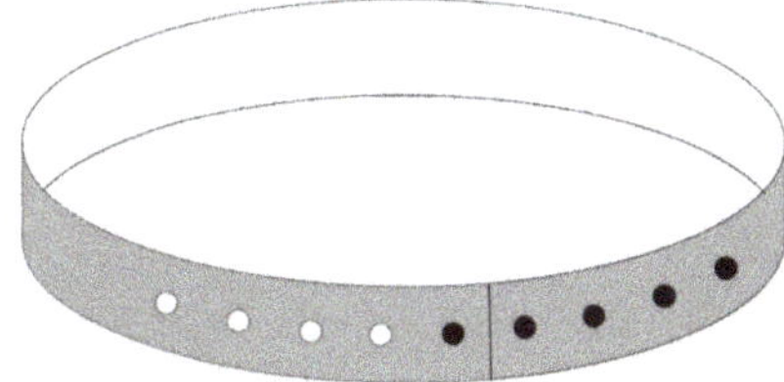

(A) 4 cm
(B) 8 cm
(C) 10 cm
(D) 16 cm
(E) 20 cm

20 In an ancient language these symbols

represent the following numbers: 1, 2, 3, 4, and 5. Nobody knows which symbol represents which number.
We know that:

Which symbol represents the number 3?

(A) 

(B) 

(C) 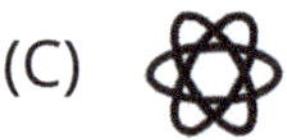

(D) 

(E) 

---

21 The hexagonal stained-glass tile is flipped. One of the flips is shown. What does the stained-glass tile look like at the far right?

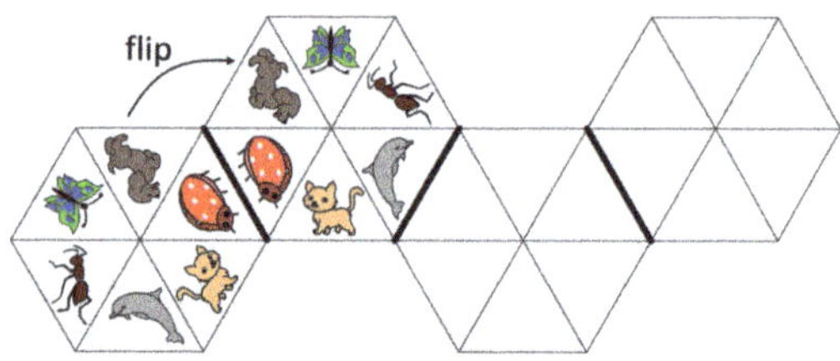

(A) 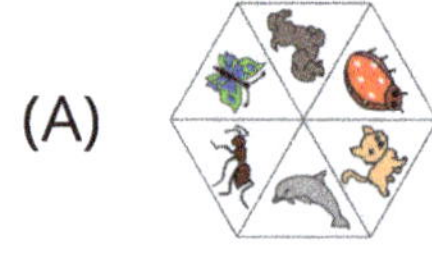

(B) 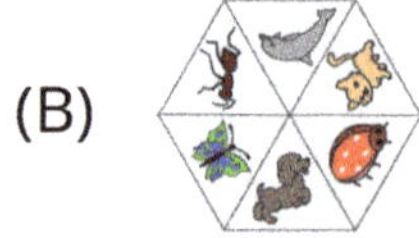

(C) 

(D) 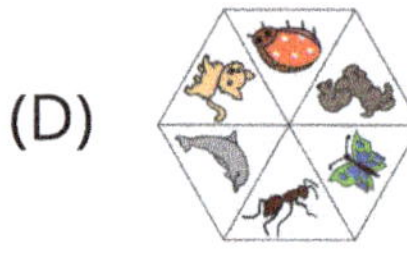

(E) 

22 The large rectangle is made up of a number of squares of various sizes. The 3 small squares each have an area of 1. What is the area of the large rectangle?

(A) 165
(B) 176
(C) 187
(D) 198
(E) 200

---

23 Leon wants to write the numbers from 1 to 7 in the grid shown. Two consecutive numbers cannot be written in two neighboring cells. Neighboring cells are those that meet at the edge or at a corner. What numbers can he write in the cell marked with the question mark?

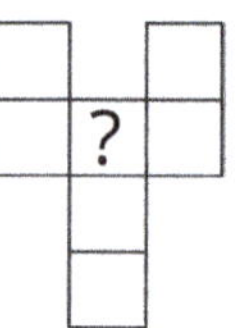

(A) all seven numbers
(B) all of the odd numbers
(C) all of the even numbers
(D) only the number 4
(E) only the numbers 1 or 7

---

24 To defeat a dragon, Matthias has to cut off all the dragon's heads. If he can cut off 3 of the dragon's heads, one new head immediately grows. Matthias defeats the dragon by cutting off 13 heads in total. How many heads did the dragon have at the beginning?

(A) 8
(B) 9
(C) 10
(D) 11
(E) 12

2020

3 Points Each

1 A mushroom grows a little every day. Mary takes a picture of the mushroom each day from Monday to Friday. Which of these pictures was taken on Tuesday?

(A) 

(B) 

(C) 

(D) 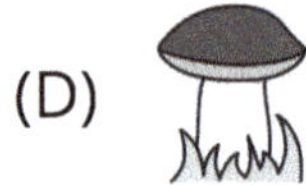

(E) 

2 Which piece completes the pattern?

(A) 

(B) 

(C) 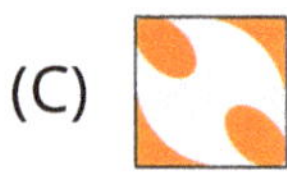

(D) 

(E) 

3 Tyler shades all the squares in the grid where the result is 20. Which shape does he get?

$16+4$	$19+1$	$28-8$
2×10	$16-4$	7×3

(A) 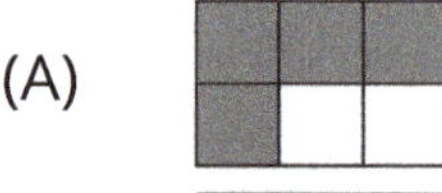

(B) 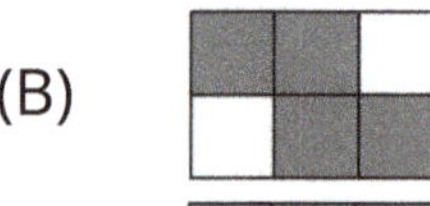

(C) 

(D)

(E)

4 Which of the following figures has the largest part shaded?

(A) 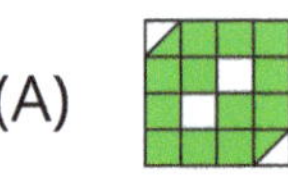

(B) 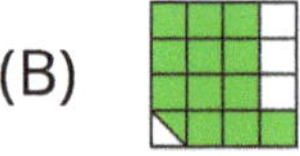

(C) 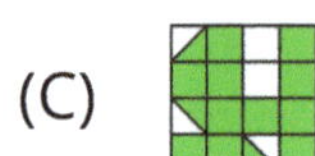

(D) 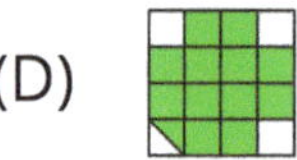

(E) 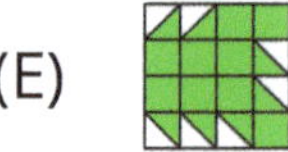

5 Which one of the figures below can you make using only these shapes:

(A) 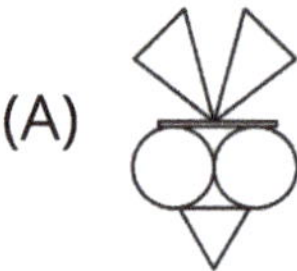

(B) 

(C) 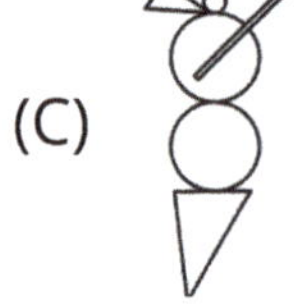

(D) 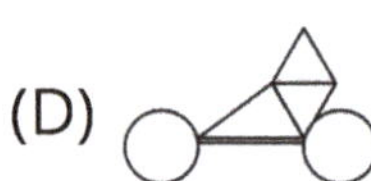

(E) 

6 Elli draws the big square (shown in the picture) with chalk on the pavement. She starts at the square marked with the number 1 and begins jumping. Each time she jumps, she always jumps to a number that is 3 more than the number she is standing on. What is the largest number Elli can jump to?

1	5	8	11
4	7	10	14
24	23	13	18
21	19	16	20

(A) 11
(B) 14
(C) 18
(D) 19
(E) 24

7 Jorge glues these 6 stickers to the faces of a cube:

The pictures show the cube in two positions.

 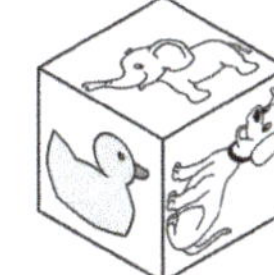

Which sticker is on the face opposite the duck?

(A) 

(B) 

(C) 

(D) 

(E) 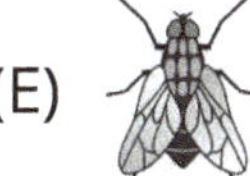

8 Casper has the following 7 pieces:

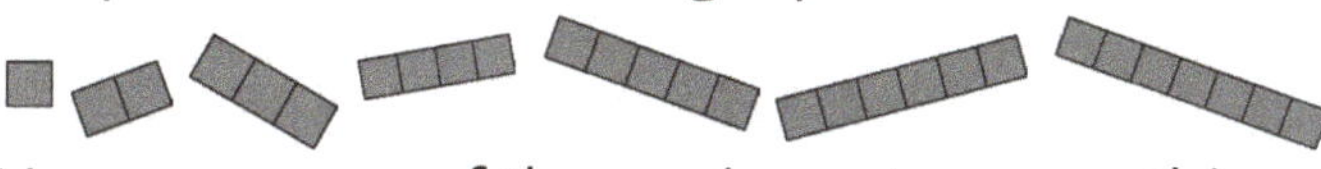

He uses some of these pieces to cover this grid: completely without overlap. He uses as many different pieces as possible. How many pieces does Casper use?

(A) 3
(B) 4
(C) 5
(D) 6
(E) 7

4 Points Each

9 Cindy colors each region on the pattern red, blue, or yellow. She colors regions that touch each other with different colors. She colors the outer ring (region) of the pattern red. How many regions does Cindy color red?

(A) 1
(B) 2
(C) 3
(D) 4
(E) 5

10 Loes looks at the pyramid from above. What does Loes see?

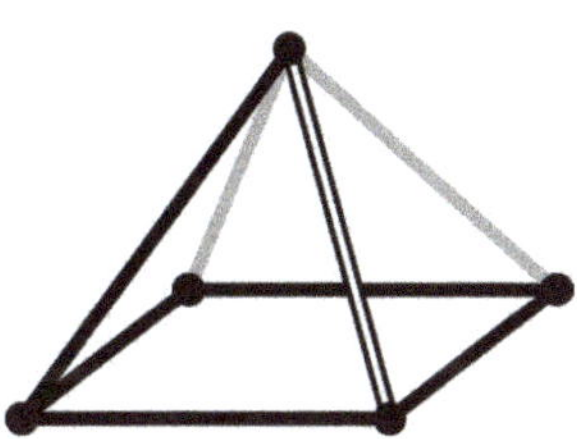

(A) 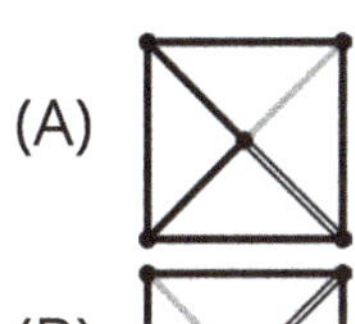

(B) 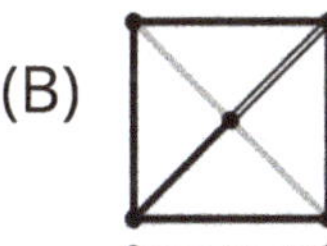

(C) 

(D) 

(E) 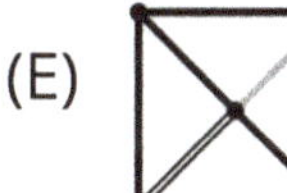

11 Dennis ties his dog 1 meter from a corner of a 7-meter by 5-meter hut as shown in the picture. He uses an 11-meter-long leash. Dennis places 5 treats as shown. How many of the treats can the dog reach?

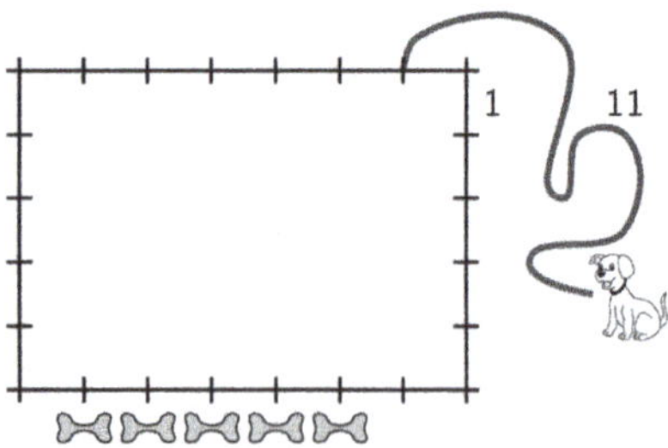

(A) 1
(B) 2
(C) 3
(D) 4
(E) 5

---

12 Lonneke builds a fence using 1-meter-long poles . The picture shows a 4-meter-long fence. How many poles does Lonneke need to build a 10-meter-long fence?

(A) 22
(B) 30
(C) 33
(D) 40
(E) 42

4 meters

---

13 Every time the kangaroo goes up 7 steps, the rabbit goes down 3 steps. On which step do they meet?

(A) 53
(B) 60
(C) 63
(D) 70
(E) 73

100
99

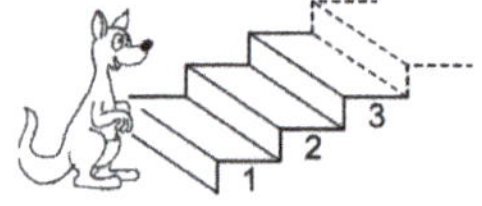

---

14 The sum of three numbers is 50. Karin subtracts a secret number from each of these three numbers. She gets 24, 13, and 7 as the results. Which of the following is one of the original three numbers?

(A) 9
(B) 11
(C) 13
(D) 17
(E) 23

---

15 Amelie wants to build a crown using 10 copies of this token:

When two tokens share a side, the corresponding numbers match. Four tokens have already been placed. Which number goes in the triangle marked with an X?

(A) 1
(B) 2
(C) 3
(D) 4
(E) 5

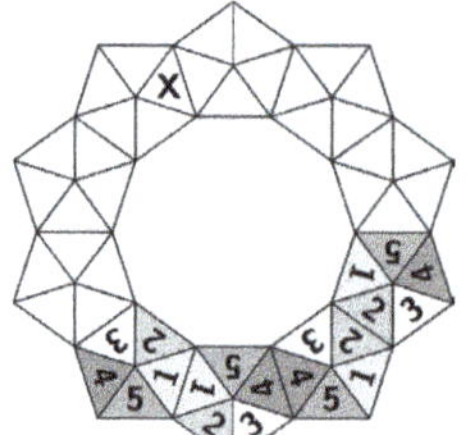

---

16 Farid has two types of sticks: short ones measuring 1 cm, and long ones measuring 3 cm. With which of the combinations below can he make a square, without breaking or overlapping the sticks?

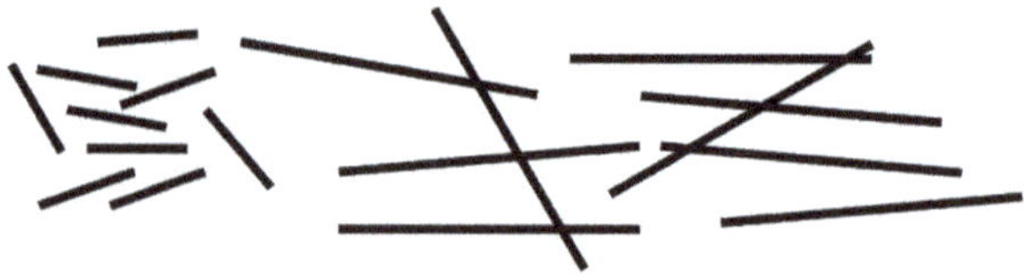

(A) 5 short and 2 long
(B) 3 short and 3 long
(C) 6 short
(D) 4 short and 2 long
(E) 6 long

5 Points Each

17 A standard die has 7 as the sum of the dots on opposite faces. The die is put on the first square as shown and then rolls towards the right. When the die gets to the last square, what is the total number of dots on the three faces marked with the question marks?

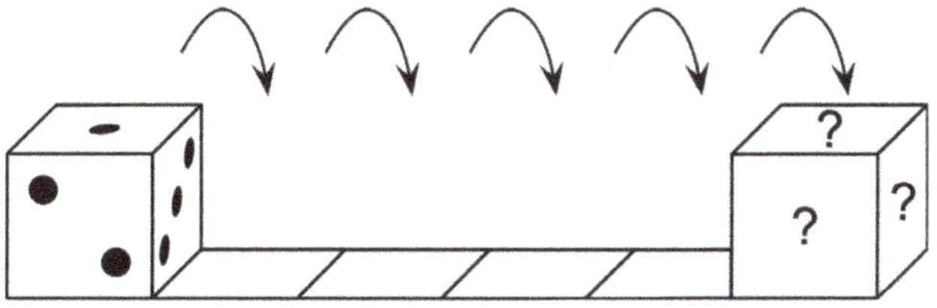

(A) 6
(B) 7
(C) 9
(D) 11
(E) 12

---

18 Six people each order one scoop of ice cream. They order 3 scoops of vanilla, 2 scoops of chocolate, and 1 scoop of lemon. They top the ice cream with 3 cherries, 2 wafers, and 1 chocolate chip. They use one topping on each scoop, and no two ice cream desserts are the same. Which of the following combinations is not possible?

(A) chocolate with a cherry
(B) vanilla with a cherry
(C) lemon with a wafer
(D) chocolate with a wafer
(E) vanilla with a chocolate chip

19 The Queen tries to find out the three names of Rumpelstiltskin's wife. She asks her:

"Are you called Adele Lilly Cleo?"
"Are you called Adele Laura Cora?"
"Are you called Abbey Laura Cleo?"

Each time exactly one name and its position are right? What is the name of Rumpelstiltskin's wife?

(A) Abbey Lilly Cora
(B) Abbey Laura Cora
(C) Adele Laura Cleo
(D) Adele Lilly Cora
(E) Abbey Laura Cleo

---

20 The teacher writes the numbers from 1 to 8 on the board. The teacher then covers the numbers with triangles, squares, and a circle. If you add the four numbers covered by the triangles, the sum is 10. If you add the three numbers covered by the squares, the sum is 20. Which number is covered by the circle?

(A) 3
(B) 4
(C) 5
(D) 6
(E) 7

21 Jane has some drawings of parrots. She wants to color only the head, tail, and wings of each parrot red, blue, or green so that all three colors are used on each picture. She colors one parrot's head red, its wings green, and its tail blue. How many more parrots can she color so that all the parrots are colored differently?

(A) 1
(B) 2
(C) 4
(D) 5
(E) 9

---

22 Several teams came to the Kangaroo summer camp. Each team has 5 or 6 members. There are 43 people in total. How many teams are at this camp?

(A) 4
(B) 6
(C) 7
(D) 8
(E) 9

---

23 Which key would it be impossible to cut into three different figures of five shaded squares?

(A) 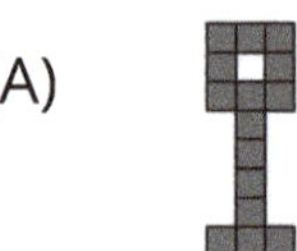

(B)

(C)

(D)

(E)

---

24 Ann replaces letters in the calculation KAN + GA – ROO with some of the numbers from 1 to 9 and then calculates the result. The same letters are replaced by the same numbers and different letters by different numbers. What is the largest possible result she can get?

(A) 925
(B) 933
(C) 939
(D) 942
(E) 948

2022

3 Points Each

1 Buzz the bee wants to reach the flower. Which set of directions will get him there?

(A) → ↓ → ↓ ↓ →
(B) ↓ ↓ → ↓ ↓
(C) → ↓ → ↓ →
(D) → → ↓ ↓ ↓
(E) ↓ → → ↓ ↓ ↓

2 Laser beams reflect in mirrors in the way shown in the first picture. At which letter will the laser beam in the second picture end?

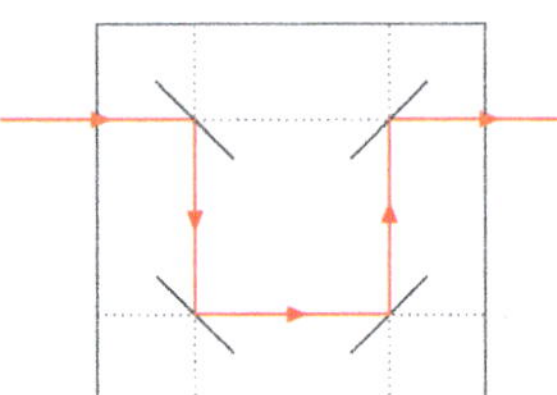

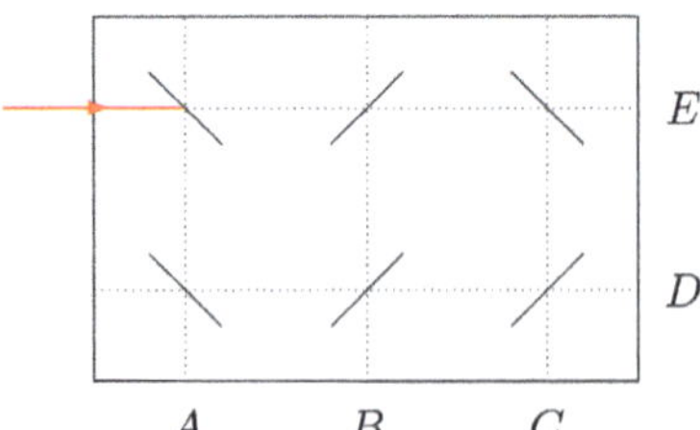

(A) *A*
(B) *B*
(C) *C*
(D) *D*
(E) *E*

3 Rossitza wants to put 2 coins in each row and in each column of the grid. Which coin does she need to move to an empty cell?

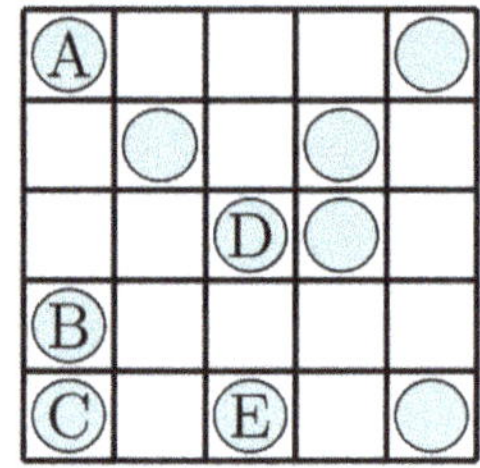

(A) A
(B) B
(C) C
(D) D
(E) E

4 What is the smallest number of boxes that Bill has to move to be able to open the TRAIN box?

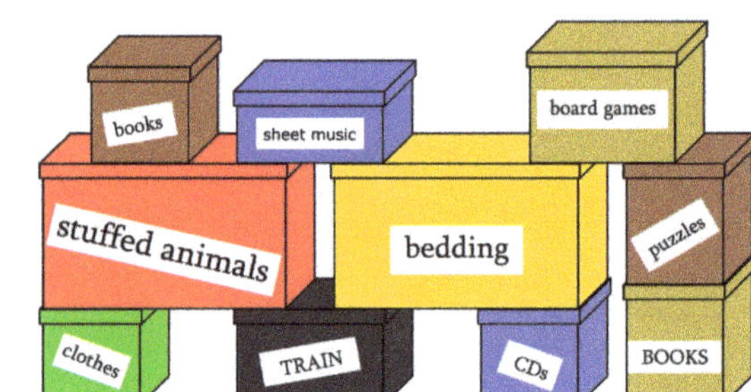

(A) 3
(B) 4
(C) 5
(D) 6
(E) 7

5 Kengu always makes one large jump followed by two small jumps on the number line, as shown in the picture. Kengu starts at 0 and ends on 16. What is the number of jumps that Kengu makes?

(A) 4
(B) 7
(C) 8
(D) 9
(E) 12

6 Anna makes a puzzle where two squares with common sides do not contain the same number. Which piece should she use to complete her puzzle?

3	2	5	4	2	1
1	4	3	1	3	4
2	5		5	2	1
4	1				3
3	2	4	2	5	2
4	1	3	1	3	4

(A) 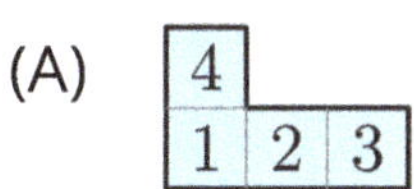

(B) 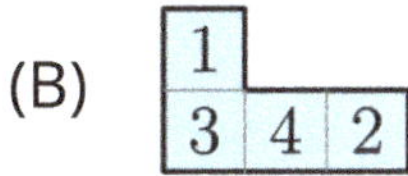

(C)
2
4 1 3

(D)
2
3 1 4

(E)
3
2 1 4

7 Which two numbers can be written in the two boxes to make the statement correct?

2022 + ☐ = 2020 + ☐

(A) 3 and 5
(B) 4 and 1
(C) 3 and 4
(D) 7 and 2
(E) 9 and 8

8 John builds the tower shown.

What will he see if he looks at his tower from above?

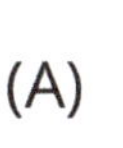 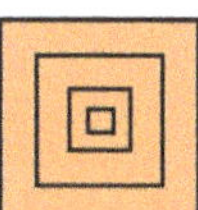

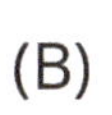 

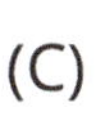 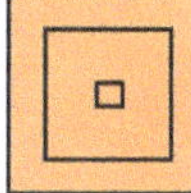

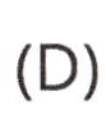 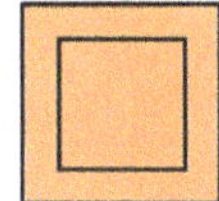

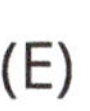 

4 Points Each

9 Five cars numbered 1, 2, 3, 4, and 5 are moving in the same direction. First, the last car (5) passes the two cars ahead of it. Next, the second to last car passes the two cars ahead of it. Finally, the middle car passes the two cars ahead of it. In what order are the cars now?

1 2 3 4 5

(A) 1, 2, 3, 5, 4
(B) 2, 1, 3, 5, 4
(C) 2, 1, 5, 3, 4
(D) 3, 1, 4, 2, 5
(E) 4, 1, 2, 5, 3

10 The ages of a family of kangaroos are 2, 4, 5, 6, 8, and 10 years. The sum of the ages of four of them is 22 years. What are the ages of the other two kangaroos?

(A) 2 and 8
(B) 4 and 5
(C) 5 and 8
(D) 6 and 8
(E) 6 and 10

---

11 During my vacation I sent the five postcards shown below to my friends.

- There are no ducks on Mike's card.
- Cara's card has the sun on it.
- There are exactly two living creatures on Paula's card.
- Lexi's card has a dog on it.
- There are kangaroos on Heather's card.

Which card did Mike get?

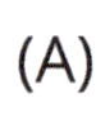 (A) 

 (B) 

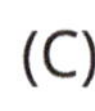 (C) 

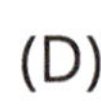 (D) 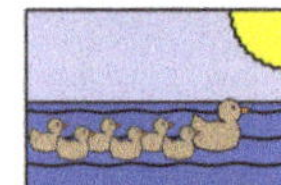

(E) 

12 George wanted the sum of the three numbers in each row and in each column of the grid to be the same. He made one mistake. Which number must he correct?

9	1	5
3	7	6
4	7	4

(A) 1
(B) 3
(C) one of the 4s
(D) 5
(E) one of the 7s

---

13 Aladdin has a square carpet. There is the same number of dots, arranged in two lines, along each of the four sides of the carpet. Unfortunately, one side of the carpet got folded over. How many dots are there on Aladdin's carpet?

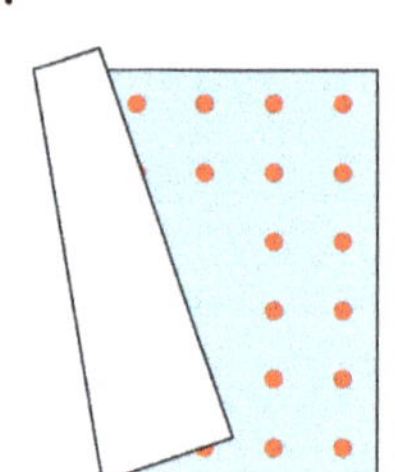

(A) 48
(B) 44
(C) 40
(D) 36
(E) 32

---

14 Joanna folds the number square twice as shown. Then she punches a hole through the black spot shown by the arrow. Which numbers does she punch through?

1	2	3	4	5	6
7	8	9	10	11	12
13	14	15	16	17	18
19	20	21	22	23	24
25	26	27	28	29	30
31	32	33	34	35	36

(A) 8, 11, 26, 29
(B) 14, 17, 20, 23
(C) 15, 16, 21, 22
(D) 14, 16, 21, 23
(E) 15, 17, 20, 22

15 The students in a class sit in rows. There is the same number of students in each row. There are 2 rows of students in front of Robert and 1 row of students behind him. In his row, there are 3 students on his left and 5 students on his right. How many students are there in this class?

(A) 10
(B) 17
(C) 18
(D) 27
(E) 36

---

16 The cube in the picture is built using the three kinds of wooden blocks shown. How many white wooden blocks are used?

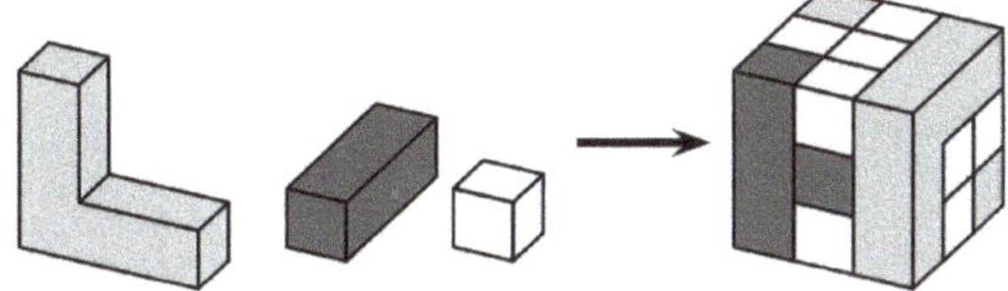

(A) 8
(B) 11
(C) 13
(D) 16
(E) 19

5 Points Each

17 Wanda chose a few of the following shapes

   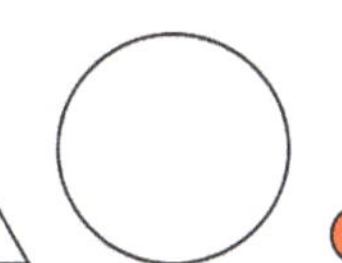

and said, "Among the shapes I chose, there are 2 colored shapes, 2 large shapes, and 2 round shapes." What is the smallest number of shapes that Wanda could have chosen?

(A) 2
(B) 3
(C) 4
(D) 5
(E) 6

18 Three football teams participate in a sports tournament. Each team plays the other two teams exactly once. In each game, the winner gets 3 points and the loser doesn't get any points. If the game ends in a tie, each team gets 1 point. At the end of the tournament, which number of points is it impossible for any team to have?

(A) 1
(B) 2
(C) 4
(D) 5
(E) 6

---

19 A pyramid is built out of cubes. Each cube has a side length of 10 cm. An ant climbed up and over the pyramid, as shown by the red line. What is the length of the red path the ant walked?

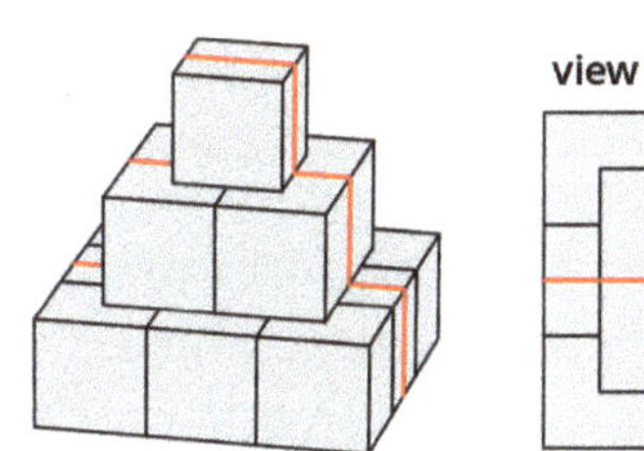

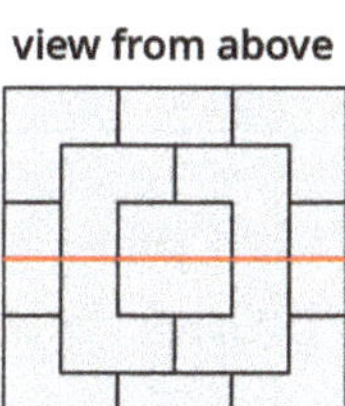

(A) 30 cm
(B) 60 cm
(C) 70 cm
(D) 80 cm
(E) 90 cm

---

20 Alma wants to put one of the pieces shown in the middle of the picture so that a child in A is able to travel to B and to E, but not to D. She can rotate the pieces. Which two pieces could she use?

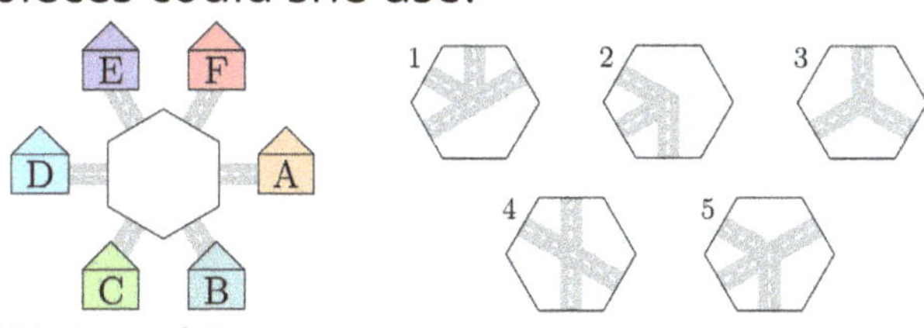

(A) 1 and 2
(B) 2 and 3
(C) 1 and 4
(D) 4 and 5
(E) 1 and 5

21 Ahmad and Zhaleh start moving from point A with the same speed, in the directions shown. Ahmad walks around the square-shaped garden and Zhaleh walks around the rectangular-shaped one. They meet again at A. What is the smallest number of laps around the square-shaped garden that Ahmad can make to meet Zhaleh there?

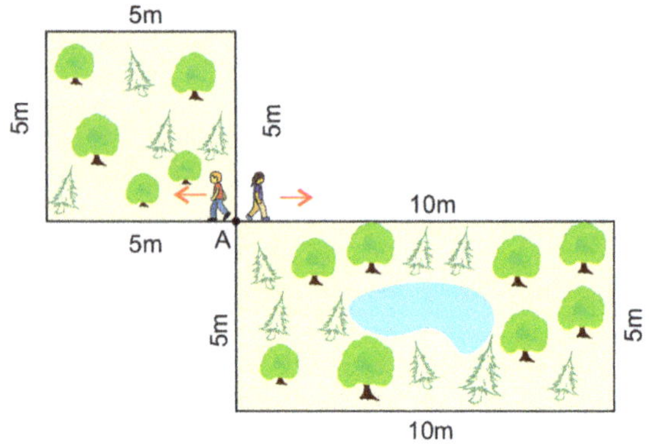

(A) 1
(B) 2
(C) 3
(D) 4
(E) 5

---

22 Five girls ate some plums. Lauren ate two plums more than Sophie. Betty ate three plums fewer than Lauren. Claire ate one plum more than Betty and three plums fewer than Alice. Which two girls ate the same number of plums?

(A) Claire and Lauren
(B) Claire and Sophie
(C) Lauren and Alice
(D) Sophie and Alice
(E) Alice and Betty

23 The small caterpillar shown in the picture curls up to sleep. What might it look like?

(A)

(B)

(C)

(D)

(E)

---

24 In the grid, the same number is hidden under the same color square. To the right of each row, the sum of the numbers hidden under the squares in that row is given. Which number is hidden under the black square?

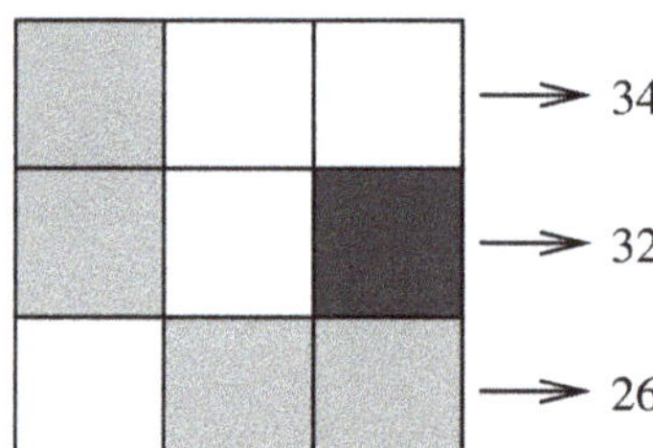

(A) 6
(B) 8
(C) 10
(D) 12
(E) 14

Part II
Solutions

1998

3 Point Solutions

1 **(C) 9**
Each kangaroo had four legs, two ears, and one tail, so Bob counted 4 + 2 + 1 = 7 parts for every kangaroo. 63 ÷ 7 = 9, so Bob saw 9 kangaroos. Check:

7 + 7 + 7 + 7 + 7 + 7 + 7 + 7 + 7
= 9 × 7 = 63

---

2 **(D) 6 × 8 + 20 ÷ (4 – 2)**
Remember the order of operations. First do what is in the parentheses. Then, from left to right, do multiplication and division, and finally do addition and subtraction, also from left to right.

(A) 6 × (8 + 20) ÷ 4 – 2 = 6 × 28 ÷ 4 – 2 =
168 ÷ 4 – 2 = 42 – 2 = 40

(B) (6 × 8 + 20 ÷ 4) – 2 = (48 + 20 ÷ 4) – 2 =
(48 + 5) – 2 = 53 – 2 = 51

(C) (6 × 8 + 20) ÷ 4 – 2 = (48 + 20) ÷ 4 – 2 =
68 ÷ 4 – 2 = 17 – 2 = 15

(D) 6 × 8 + 20 ÷ (4 – 2) = 6 × 8 + 20 ÷ 2 =
48 + 10 = 58

(E) 6 × (8 + 20 ÷ 4) – 2 = 6 × (8 + 5) – 2 =
6 × 13 – 2 = 78 – 2 = 76

3 **(C) 8**
There are two large overlapping triangles (one marked red and one marked purple in the picture) and six small triangles (marked green).

---

4 **(D) 7th**
Apartments 1, 2, 3 are on the 2nd floor; apartments 4, 5, 6 are on the 3rd floor; apartments 7, 8, 9 are on the 4th floor; apartments 10, 11, 12 are on the 5th floor; apartments 13, 14, 15 are on the 6th floor; and apartments 16, 17, 18 are on the 7th floor. Mary lives on the 7th floor.

---

5 **(C) 48**
12 × 12 × 12 = (6 × 2) × 12 × (2 × 6) =
6 × (2 × 12 × 2) × 6 = 6 × 48 × 6

---

6 **(C) 4**
A three-digit number cannot start with a zero. So, the only possible numbers are 307, 370, 703, and 730.

---

7 **(D) 28**
In the very center of the tower there are 4 vertical blocks, and around it there are 4 staircase-like structures, each made of 6 blocks. The total number of blocks is 4 + 4 × 6 = 28.

8 **(B) 54**

Work backwards. The last step was to add 18, so subtract 18 from the final number. 180 – 18 = 162. Before that, the number was multiplied by 3, so divide by 3. 162 ÷ 3 = 54. 54 was the number that was chosen at the beginning.

4 Point Solutions

9 **(C) 10 minutes**

Try different paths. Zigzagging through the middle part of the figure gives the shortest time.

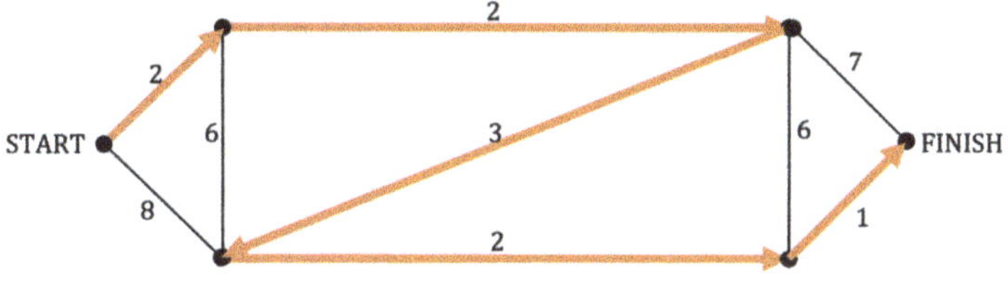

10 **(E) 13**

Possible numbers of cookies for 3 plates when 1 cookie is always left over are: 4, 7, 10, 13, 16, 19, 22, 25 , 28, ... (multiples of 3 plus 1).

Possible numbers of cookies for 4 plates when 1 cookie is always left over are: 5, 9, 13, 17, 21, 25, 29, ... (multiples of 4 plus 1).

The smallest number on both lists is 13. If Joanna divides evenly 13 cookies between 2 plates, 1 cookie is also left over.

11 **(C) 1290**

To find a solution, reverse your calculations. If something was added, subtract it; if it was subtracted, add it back. 3250 – 2000 = 1250 and 1250 + 40 = 1290.

12 **(C) Thursday**

The well is 5 meters deep. On Monday the snail goes up 2 meters but slides 1 meter down. It starts Tuesday with still 4 meters to go up, moves 2 meters up and slides 1 meter down, so on Wednesday it starts with 3 meters left to climb. Again it goes up 2 meters and slides 1 meter down. On Thursday morning it is 2 meters from the top which it climbs during the day and finally gets out of the well.

13 **(A)**

Rotate piece Z so it is upside-down, and then piece Z fits with piece (A).

14 **(B) 7**

There were 30 runners if we don't count John. They can be divided into 5 groups of equal size: 1 group that finished before him and 4 groups that finished after him. There were 6 runners in each group since 5 × 6 = 30. There was one group of 6 runners before John, so he finished in the 7th place.

15 **(D) 24**

The difference between half and one-fourth of a loaf of bread is one-fourth of a loaf of bread. Since-one fourth of a loaf costs 6 pence, the whole loaf costs 4×6 pence = 24 pence.

---

16 **(E) 3:00**

The hour hand (the thicker, shorter hand) is pointing directly at an hour mark. This only happens when the time is "on the hour." So, the minute hand is pointing at 12. Rotate the clock so that 12 is on top, and you can see that the hour hand is pointing to the third hour mark, and the time is 3:00.

5 Point Solutions

17 **(B) 7**

First, look at this part of the pyramid:

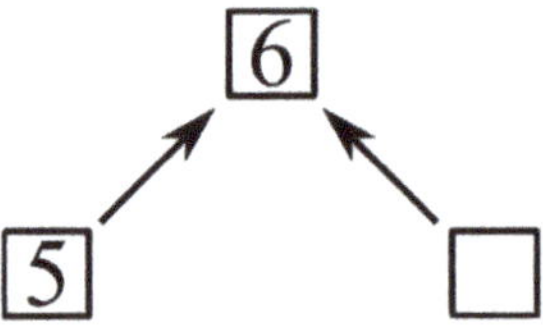

The pattern shows us to add the two lower numbers together and then divide by 2 to get the top number. In this blank box, we have to write a number such that 5 plus this number is the double of 6. $12 \div 2 = 6$, and $5 + 7 = 12$, so 7 is that number.

Now that we know that there is a 7 in the middle bottom box, let's look at this part:

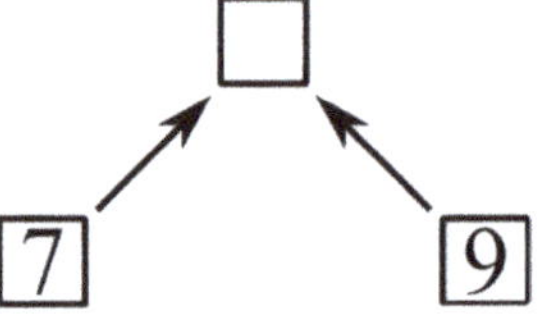

In the blank box we have to write 8, since

$$\frac{(7 + 9)}{2} = 8$$

At the top of the pyramid we have to write

$$\frac{(6 + 8)}{2}$$ which is 7.

Here is the full pyramid:

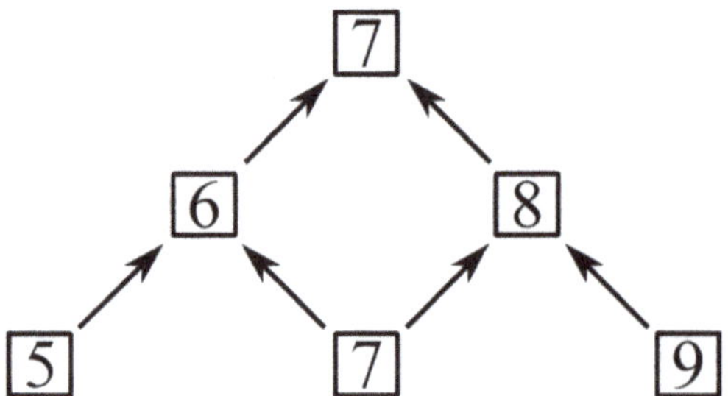

18 **(D) 7**

If there were 2 or more red balls, then the number of white balls would be at least 7 × 2 = 14, and the number of red and white balls would already be more than 15. Thus, there is only 1 red ball and 7 × 1 = 7 white balls. 1 + 7 + 7 = 15, so the other 7 balls are black.

---

19 **(E) 1 dollar and 10 cents**

Notice that if Paul added 80 cents to the 30 cents he had left, he would have enough money for another serving of ice cream. So, one serving costs 80 + 30 = 110 cents, which is equal to 1 dollar and 10 cents.

---

20 **(D) 840**

1 meter = 100 cm and 20 × 5 cm = 100 cm, so the size of the chessboard is 20 × 20 matches.

We will count vertical matches in each row and horizontal matches in each column.

If you have a row of 20 adjacent squares, then the number of vertical matches is 20 + 1 = 21.

There are 20 rows, so the number of all vertical matches is 20 × 21.

We can count the number of horizontal matches in the same way, so 20 × 21 is the number of all horizontal matches. The number of all the matches is 2 × 20 × 21 = 840.

21 **(B) 15**

The numbers, in their natural order, are: 104, 113, 122, 131, 140; 203, 212, 221, 230; 302, 311, 320; 401, 410, and 500. There are 15 numbers on the list.

---

22 **(B) 6 years**

Look at the line below showing the age differences from youngest to oldest. Each girl is represented by her initial.

+1 +3 +2

D B A C

1 + 3 + 2 = 6, so Cali is 6 years older than Dorothy.

---

23 **(B) 5**

My team received 64 points for ties and wins. The team earned 7 points for 7 ties and 57 points for a number of wins (7 + 57 = 64). The team won 19 games since 19 × 3 = 57 and 7 games were ties, so my team lost 31− (7 + 19) = 31 −26 = 5 games.

---

24 **(D) 95**

The one-digit numbers (1 to 9) take up positions from 1 to 9. Then the two-digit numbers start in position 10, and each takes up 2 positions.

12345678910111213141516...

There are 90 two-digit natural numbers (10 to 99 inclusive), which is 180 digits, which would make the last 9 of 99 be in the 189th positions.

...90919293949596979899

Going back 9 positions to 180 we are at the 9 of 95. So, the natural number we are looking for is 95.

2000

3 Point Solutions

1 **(B) 15 minutes**
All 10 candles will stay lit for 15 minutes since they are lit at the same time.

2 **(C) 40**
Look at the timeline below. The little kangaroo will take the last pill 20 minutes + 20 minutes = 40 minutes after taking the first pill.

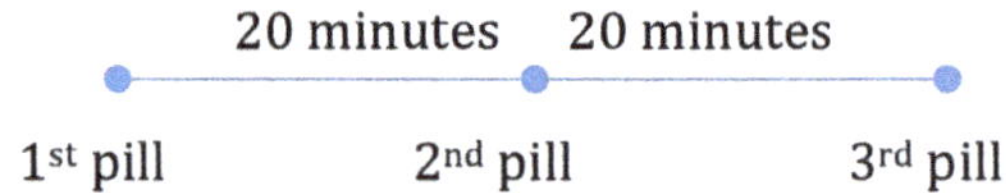

3 **(D) 222**

Number	112	209	312	222	211
Product of digits	2	0	6	8	2
Sum of digits	4	11	6	6	4

The product of the digits is greater than the sum of the digits only for the number 222.

4 **(B) on the 3rd floor**
Gavel has to climb one flight of stairs to get to the second floor of the building and Pavel has to climb two flights of stairs: one to get to the second floor, and another one to get from there to the third floor. So, Pavel lives on the 3rd floor.

5 **(B) 30 cents**
$4.50 is four dollars and 50 cents. Each candy bar costs 90 cents, so four candy bars cost 3.60 dollars. Subtracting from the total price of 4.50 dollars, we have 90 cents remaining for the three lollipops, so one lollipop costs 30 cents.

6 **(C) 3**
160 = 55 + 55 + 50, so at least 3 buses are needed to seat 160 people.

7 **(B) 24 minutes**
A big square has the area four times greater than the area of a small square. The blue arrows point to matching segments, so the perimeter of the small square is equal to the two sides (marked by green arrows) of the big square. Hence, the perimeter of the big square is twice the perimeter of the small square. Thus, the time to walk around the bigger square plaza is twice the time it takes to walk around the smaller plaza. 12 minutes are needed to walk around the smaller plaza, so 2 × 12 minutes = 24 minutes are needed to walk around the bigger plaza.

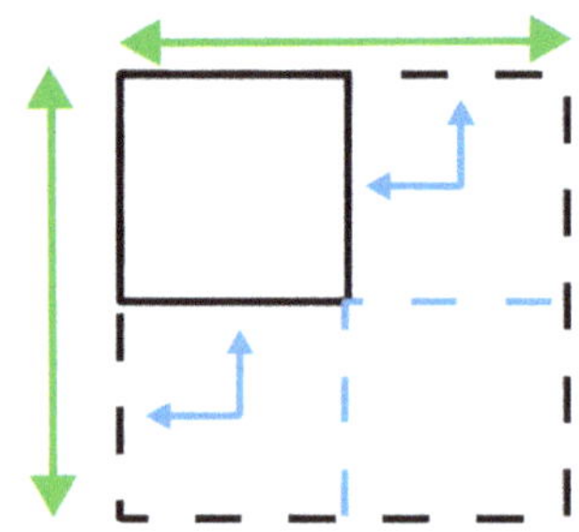

8 **(C) 9:30**
Every bus needs 2 hours to drive from Zakopane to the Krakow airport since 2 hr × 60 km per hour = 120 km, which is the distance from Zakopane to the Krakow airport. To be at the airport at 11:30, the bus had to leave Zakopane no later than 9:30.

4 Point Solutions

9 **(B) 4**

Kathy eats two bowls of ice cream during the time in which Betty eats three bowls of ice cream. So for each 5 bowls of ice cream the girls eat, Kathy eats 2 bowls. If both girls ate $10 = 2 \times 5$ bowls of ice cream, then Kathy ate $4 = 2 \times 2$ bowls of ice cream.

10 **(D) 4, 9, 2, 5**

To have the smallest possible three-digit number, select the smallest digit from 49215, which is 1, for the hundreds digit. For the tens digit, select the smallest digit from 50 which is 0, so the ones digit is 8. The removed digits are 4, 9, 2, 5.

11 **(B) 12**

However many apples Aria takes out from her basket is exactly how many apples will remain in Zoe's basket, so 12 apples were left in both baskets altogether.

12 **(D) 33**

The boys were in the seventh row from the front and the fifth row from the back, so there were six rows before them and four rows behind them. Thus, there are 11 rows with three students in each row. Hence, $11 \times 3 = 33$ students went to the museum.

13 **(B) 4**

If 5 (or more) cats were playing moms and each mom had 2 (or more) kids, then the number of all cats would be more than 14 since $5 + 5 \times 2 = 15$. The greatest possible number of cat-moms in the play is 4 as shown below.

14 **(C) 4**

According to the second scale:

6 plums, one apple, and one pear can be removed from the first scale:

and this scale will stay in balance.

Here is the new first scale:

It has two identical groups of plums on the left side and two apples on the right side, so

stays in balance. This allows us to remove two plums and one apple from the second scale and gives us the solution:

15 **(C) Mr. Jack**

Figures (B), (D), and (E) have the same perimeters as (A) as shown by matching arrows (blue is used for horizontal segments and green is used for vertical segments). For (C) vertical matchings are shown but the extra horizontal segments of (C) have no match with horizontal segments of the outer square, so Mr. Jack has the longest fence.

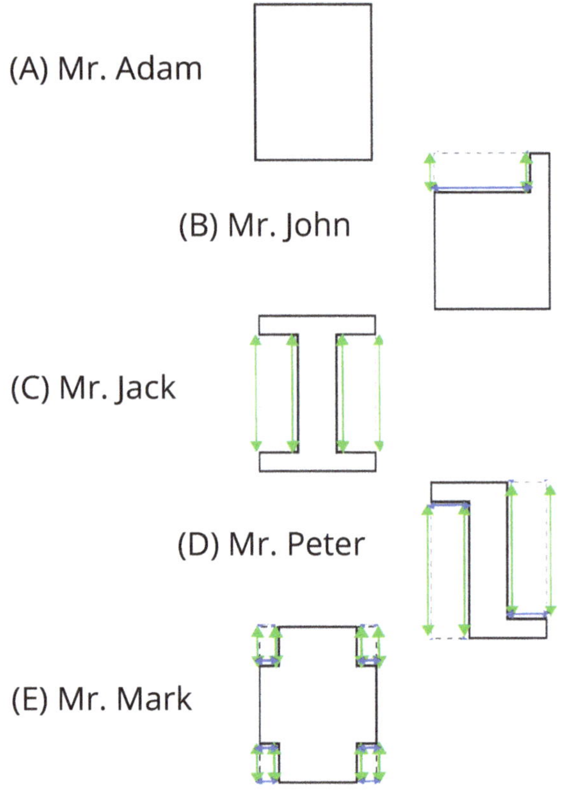

---

16 **(C) 6**

The figure on the left weighs 300 grams and consists of 10 blocks, so each block weighs 30 grams. Both figures weigh 900 grams and use identical blocks, so the second figure weighs 600 grams and consists of 20 blocks. Only 14 blocks of the second figure are visible, so 6 blocks are not shown.

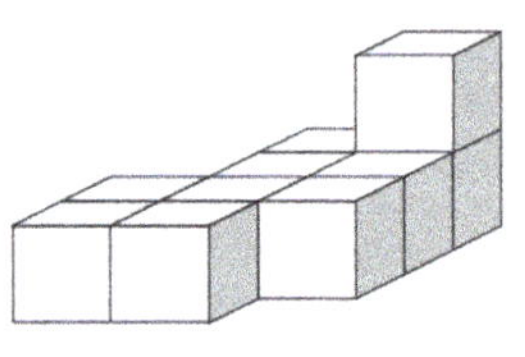

10 visible blocks

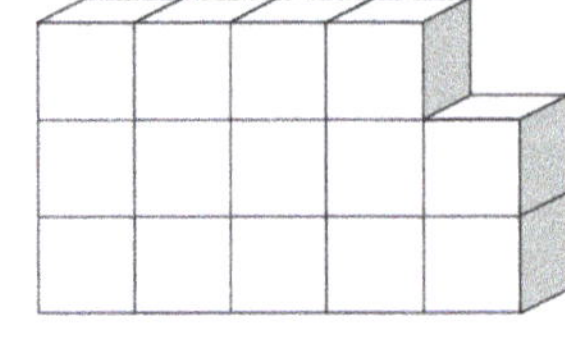

14 visible blocks

5 Point Solutions

17 **(B) 12**

6 hens eat 8 cups of grain in 3 days, so 3 hens will eat 4 cups of grain in 3 days. Hence, in 9 days, these 3 hens will eat 4 × 3 = 12 cups of grain.

---

18 **(B) 240 cm**

As shown in the picture, there are four pieces of ribbon in all. Two of them (shown in blue), each made of four 10 cm segments, give a total length of 80 cm (2 × 4 × 10 = 80). The other two (shown in orange) have two 10 cm segments and two 30 cm segments each, giving a total length of 160 cm [2 × (2 × 10 + 2 × 30) = 2 × 80 = 160]. This gives a total length of 80 cm + 160 cm = 240 cm.

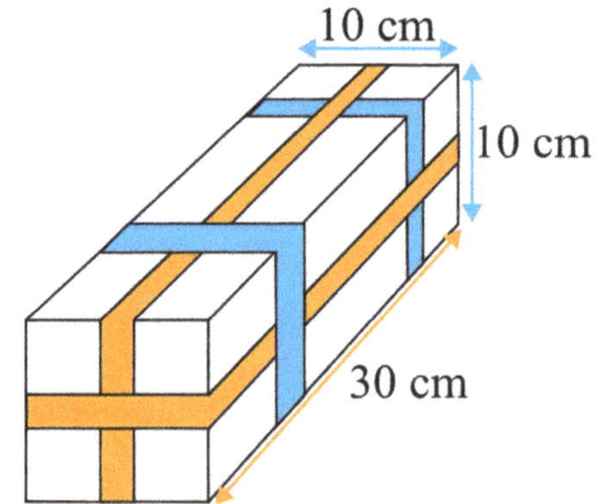

19 **(E) 2**

A number line will explain the relationship between ages of the two kangaroos. The age of the youngest kangaroo is represented by the orange arrow. To multiply it by 5, add 4 green arrows of the same size to it.

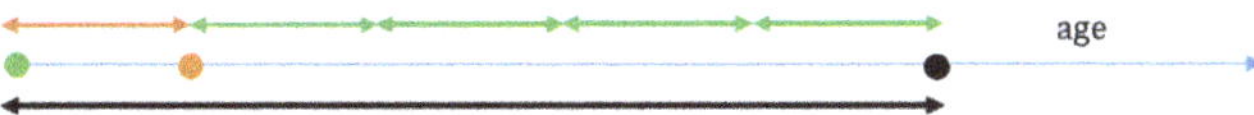

The black arrow represents the age of the oldest kangaroo. The difference between the ages is represented by the four green arrows, so it is 4 × the age of the youngest kangaroo. All three kangaroos were born 4 years apart, so the oldest kangaroo is 8 years older than the youngest one. This difference between their ages is always 8 years, so 8 = 4 × the age of the youngest one. Hence, right now, the youngest kangaroo is 2 years old.

---

20 **(C) 6 hours**

If we were to turn the clock's dial so that it shows 12:00 on the first clock (marked in red), and turn the dial on the second clock the same way, so that the minute hand is at 12 (also marked in red), then the second clock would show 6:00 as the time. Thus, the time difference is 6 hours.

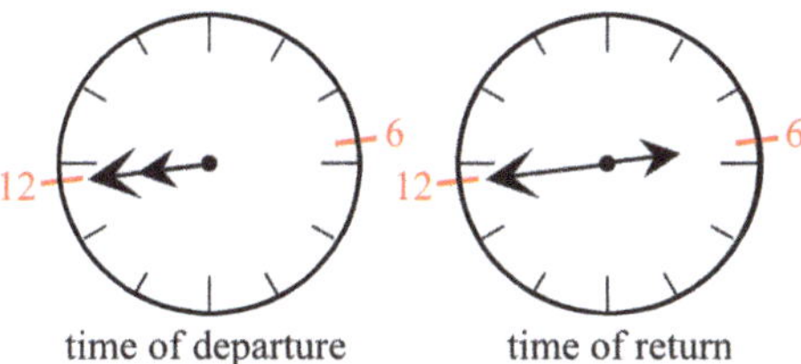

Note: The hour hand and the minute hand on the watch overlap every 1 hour, 5 minutes, and 5/11 of a minute.

21 **(C) 6**

If we add the original number and its half, then we get $\frac{3}{2}$ of the original number. If we double the original number, then we get $\frac{4}{2}$ of the original number. The difference between $\frac{4}{2}$ and $\frac{3}{2}$ is $\frac{1}{2}$ so half of the original number is 3. Therefore, the original number is 6.

---

22 **(A) 1**

The faces which are touching between the middle and top dice both must show the same number, which cannot be 6, 5, or 4 visible on the top die and cannot be 2 or 3 visible on the middle die, so it must be 1. Since the dice are identical, each die has two opposite faces with 1 and 6 dots. For the middle die, the face opposite the face with 1 dot must have 6 dots, so the top face of the lowest die must also have 6 dots. Therefore, the bottom of the lowest die has 1 dot.

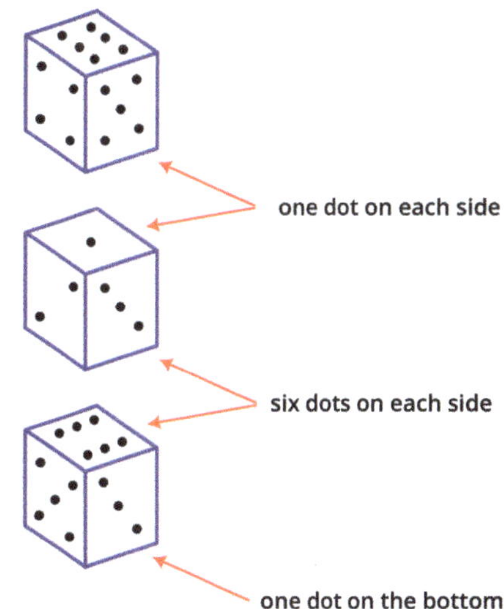

23 (E) There is no such point; this is impossible.

There is only one path connected to point A, so A is either the starting or the ending point of a drawing that is done without lifting the pencil from the paper or going over the same line twice. If A is the starting point, K will need the be ending point, and if A is the ending point, K will have to be the starting point, because there are two other paths that join at K that make the rest of the picture. There is another place where three paths join, which is point C. The only way we can draw three paths coming from one point is if we make it the starting or ending point, so that after coming to the point the first time and moving away from it, we can come back to the point and not have to leave it by a path that has been drawn already (starting at point C means moving away from it, coming back, and moving away again). However, if point C were the starting or ending point, then he could not draw the segment from K to A without going over that line twice. Since this means that the picture cannot be drawn this way, Pete has no starting point to draw the picture of a kangaroo shown without lifting his pencil from the paper and without going over the same line twice.

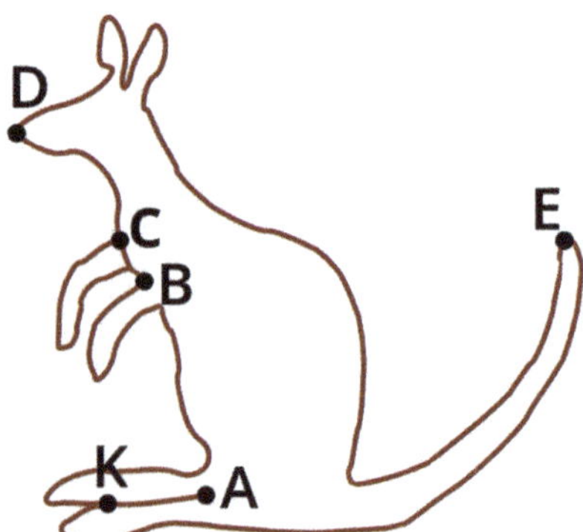

24 (A) 80 cm

The ball reached a height of 320 cm after the second bounce, so it must have reached a height of 320 cm ÷ 2 = 160 cm after the first bounce, and therefore would have been initially dropped from a height of 160 cm ÷ 2 = 80 cm.

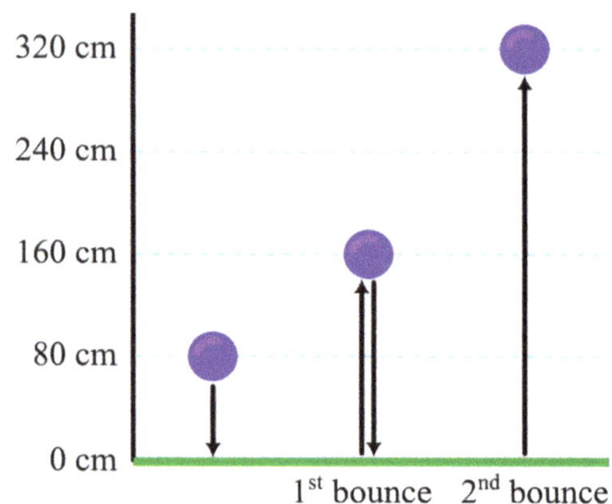

2002

3 Point Solutions

1 **(B)** 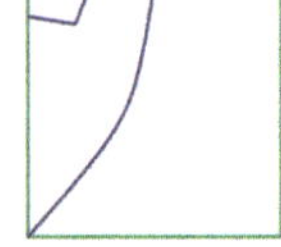

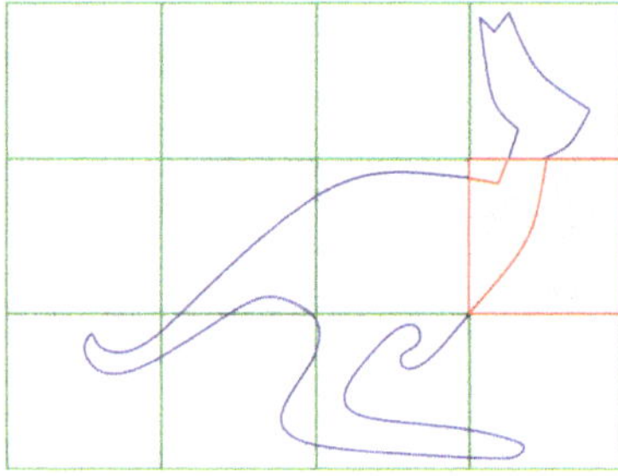

---

2 **(C) 4**

$$2 + 2 - 2 + 2 - 2 + 2 - 2 + 2 - 2 + 2 =$$
$$= 2 + \quad 0 \quad + \quad 0 \quad + \quad 0 \quad + \quad 0 \quad + 2 = 2 + 2 = 4$$

---

3 **(D) 23**
Just add the numbers of gifts Andrzej received: 3 + 4 + 3 + 10 + 2 + 1 = 23.

---

4 **(D)** 

The picture shows the figures that can be found in the square.

---

5 **(D) Joanna**
Lena and Suzie were born in the same month, so they were born in March. Gina and Suzie were born on the same day of the month, so their birthdays fall on the 20th. This leaves Joanna as the girl who was born on May 17th.

---

6 **(C) 4,200**
There are 60 minutes in an hour, so during one hour the human heart beats on average 60 × 70 = 4200 times.

---

7 **(A) 14 cm**
The lengths of two of the sides of the square *ABCD* and the longer sides of rectangle *ATMD* are equal, and the difference in lengths of the other two sides of those figures is 10 cm – 3 cm = 7 cm. The difference between the sum of the lengths of all the sides of the square and the sum of the lengths of all the sides of the rectangle *ATMD* is 2 × 7 cm = 14 cm.

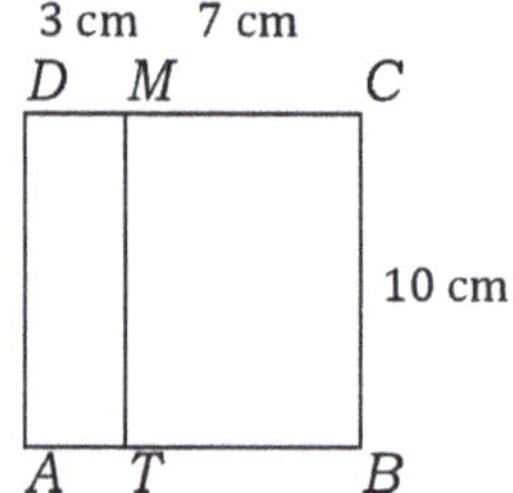

---

8 **(B)** 

The pictures below show fold lines and edges.

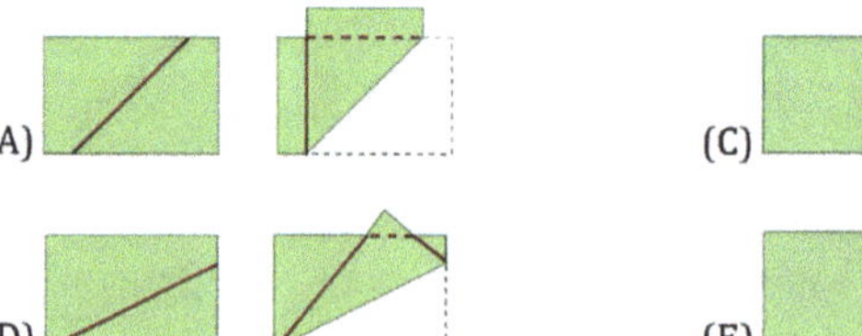

When folding any piece of paper, the result is a shape completely on one side of the folding line, so the red lines (shown to the left) cannot be the folding lines of (B). The purple lines potentially could be the folding lines of (B) but unfolding along any of the purple lines does not produce a rectangle.

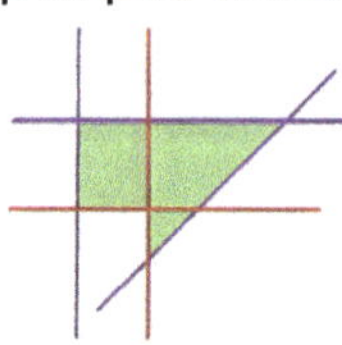

4 Point Solutions

9 **(C) 8**
The digit 2 shows up 8 times, in numbers: 2, 12, 20, 21, 22 (2 used twice), 23, and 24.

---

10 **(E) 6**
The scale is balanced when there are 2 melons on the one side and 6 oranges plus 1 melon on the other. By removing one melon from each side, we can see that the weight of 6 oranges is equal to the weight of 1 melon.

---

11 **(C)** 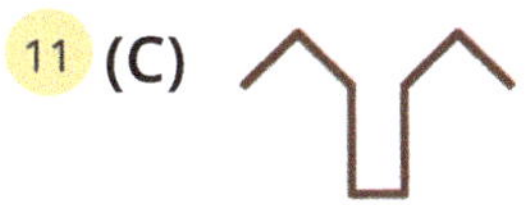

The picture shows where (A), (B), (D) and (E) can be found.

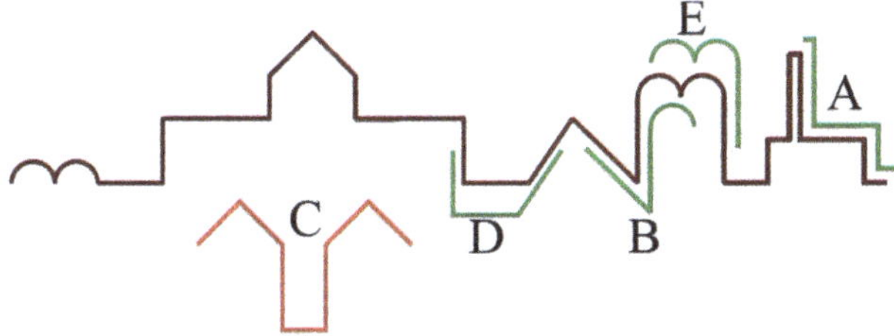

---

12 **(A) 3**
The smallest two-digit number is 10 and the greatest one-digit number is 9. The sum of 10 and 17 is 27. When 27 is divided by 9 the result is 3.

13 **(E)**

The symbols represent
60 + 60 + 1 + 1 + 1 + 1 = 120 + 4 = 124.

---

14 **(B)**

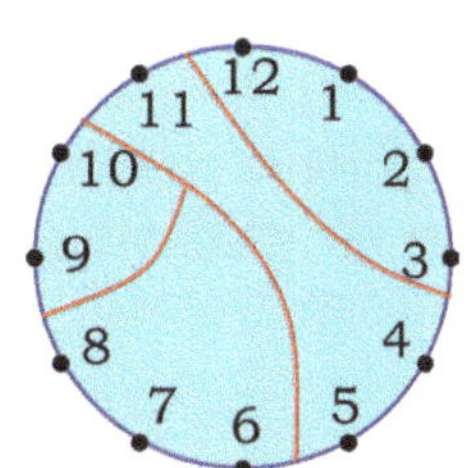

1 + 2 + 3 + 4 + 5 + 6 + 7 + 8 + 9 + 10 + 11 + 12 = 78, so the sum of all the numbers on the face of the clock is 78. The only four consecutive numbers for which the sum is 78 are 18, 19, 20, and 21. In picture (B), the sums of the numbers in each part of the clock are: 12 + 1 + 2 + 3 = 18; 10 + 9 = 19; 11 + 4 + 5 = 20; 6 + 7 + 8 = 21. For each of the other clocks we can find a part with the sum of its numbers less than 18.

15 **(C) 15**
Klara built a triangle using 6 × 3 = 18 matches. There were 60 – 18 = 42 matches left for Zoe.

The length of one of the sides of the rectangle is equal to 6 matches. So, Zoe used 6 + 6 = 12 matches to build that side and the one opposite. The number of matches left for the other two sides is 42 – 12 = 30. So, each of those sides is made out of half of the 30 matches, which is 15.

---

16 **(E) Miki and Niki finished at the same time.**
The lengths of the routes taken by the kangaroos are: Miki, 18 units; Niki, 18 units; and Oki, 17 units. Because they were jumping with equal speeds, Miki and Niki finished at the same time because their routes were the same length.

5 Point Solutions

17 **(A) Oliver has a dog.**
Nate, who has a pet with fur but doesn't like cats, has to have a dog. Because Nate owns a dog, the sentence "Oliver has a dog" is not true. Actually, Oliver owns a cat, and Nate has the dog. The other sentences are true.

---

18 **(B) at 7:20**
Mary needs 37 minutes to get to school so Zoe needs 37 – 12 = 25 minutes. Since Zoe arrived at school at 7:45, she had to leave her house 25 minutes earlier, which means she left at 7:20.

---

19 **(D) 18**

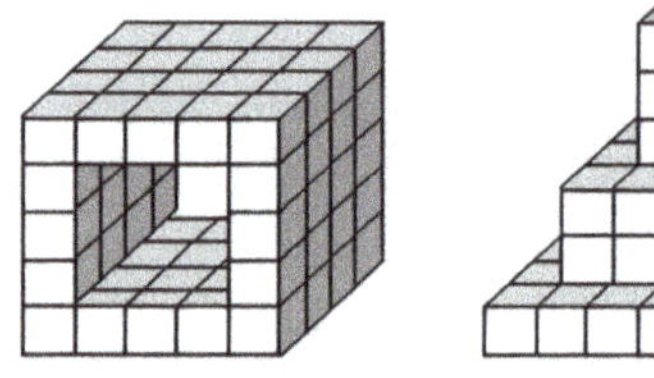

Robert made a tunnel using

5 × 5 × 4 – 3 × 3 × 4 = 100 – 36 = 64 cubes.

This also means that there were 64 cubes left at this point. He made the pyramid out of

5 × 5 + 2 (3 × 3) + 3 = 25 + 18 + 3 = 46 cubes.

Now there were 64 – 46 = 18 cubes left.

20 **(E) 11**
The green segment below represents the daughter's age. The orange segment represents the period of time after which her mom's future age is 3 times the daughter's future age.

The daughter's future age is represented by the interval with both colors and the mom's future age is represented by 3 such intervals. The difference between the future ages is still 28 since the age difference is always the same. This age difference is twice the daughter's future age. Thus, half of the age difference is 28 ÷ 2 = 14, so the daughter's future age is 14. She is 3 years old now, so the time period from now to this future point in time is 14 – 3 = 11 years.

---

21 **(C) 8**
The trio consists of a violinist, a pianist, and a drummer. A violinist alone can be selected in 2 ways. A violinist and a pianist can be selected in 2 × 2 ways since each violinist can be joined by one of 2 pianists. A whole trio is created when a violinist and a pianist are joined by one of 2 drummers, so the conductor can create 2 × 2 × 2 = 8 trios.

22 **(D) 21**
If 4 medals can be made from 4 plates, then from 16 plates we can cut out 16 medals. The remaining material is enough for 4 more plates. From those 4 plates, 4 medals can be cut out and still the remaining material is enough for 1 more plate. From that 1 plate we can have 1 more medal, therefore there are as many as 16 + 4 + 1 = 21 medals which can be made from 16 plates.

---

23 **(E) 10th**
In addition to Tom, there were 27 other students in the math competition. Since there were twice as many students with fewer points than those with more points than Tom, we can divide 27 into 3 groups, which gives 9 students in each group. Two of the groups are the students who got fewer points than Tom, and one group is the students who got more points. So, there were 9 students who got more points than Tom. Therefore, Tom finished that competition in 10th place.

---

24 **(D) after 21 km**
We cannot increase just the ones digit since it is already 9. We have to change the last two digits without using 7 or 8. The smallest such two-digit number is 90, so the odometer reading has to be 187590. To reach it the car has to travel 187590 − 187569 = 21 km.

2004

3 Point Solutions

1 **(E) 10,015**
A simple way to find the sum is to notice that 2001 + 2002 + 2003 + 2004 + 2005 = = 2000 + 2000 + 2000 + 2000 + 2000 + 1 + 2 + 3 + 4 + 5 = 10,000 + 15 = 10,015.

---

2 **(A) 4 years**
The difference in age between Mark and his sister is always the same. Today Mark is 9 years old, his sister is 9 – 4 = 5 years old, and the difference between their ages is still 4 years.

---

3 **(C) 6 km**
The new straight horizontal path on the bottom is equal to the path CD, so the only extra distance to travel are the two paths straight down from C and D. Thus, the road including the detour is 3 km × 2 = 6 km longer than the original path.

---

4 **(C) 669 – 391**
671 – 389 = 282 and the other differences are (in order): 771 – 489 = 282, 681 – 399 = 282, 669 – 391 = 278, 1871 – 1589 = 282, and 600 – 318 = 282.

Only (C) 669 – 391 = 278 is not identical to the original difference.

---

5 **(E) 14**
5 flew away and 3 returned, so there were 2 birds that didn't return. Hence, originally there were 2 more birds than there were at the end. Thus, there were 14 birds at the very beginning.

6 **(B) 1 and 10**
Numbers 1, 2, 3, 10, 11, and 13 are in the rectangle.

Numbers 1, 4, 6, 7, 9, 10, 12, and 13, are in the circle. So, numbers 1, 10, and 13 are in both the rectangle and the circle.

However, 13 is also in the triangle, so 1 and 10 are the only numbers that are in the rectangle and the circle, but not in the triangle.

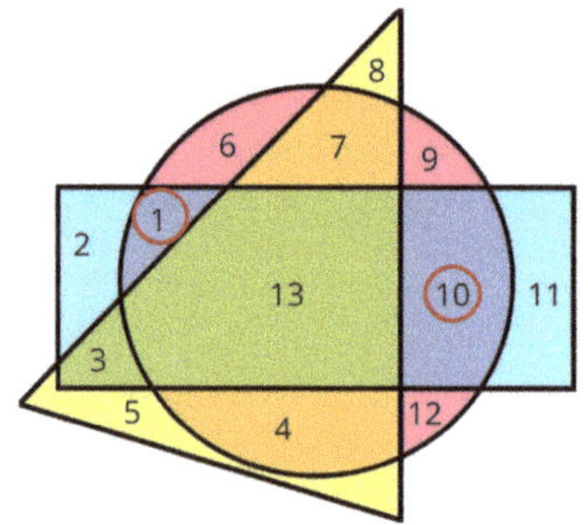

---

7 **(E) green**
The red building is next only to the blue building, which means that the red building is either building 1 or 5, and the blue building is either building 2 or 4. The blue building is between the red building and the green building. If the blue building was number 2, then the red building is number 1 and the green building is number 3. If the blue building is number 4, then the red building is number 5 and the green building is number 3. Either way, number 3 is the green building.

---

8 **(B) 3**
The figure is made of 24 squares. For the number of shaded squares to equal half the number of white squares, the number of shaded squares has to equal 8 and the number of white squares equal 16. You can get this by dividing 24 into three equal parts, with 1/3 of 24 being shaded and 2/3 of 24 not being shaded. There are already 5 squares shaded, so 3 more must be shaded.

4 Point Solutions

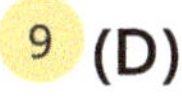
9 **(D)** 

To find a match for a completely black rectangle we have to rotate 90° either to the right (clockwise) or to the left (counterclockwise):

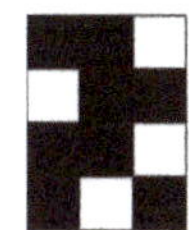

The clockwise rotation gives us:

which is not a match for any of the five options.

Below, see the pictures of the five sheets with the above rectangle placed on top of them.

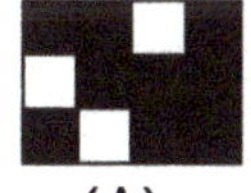  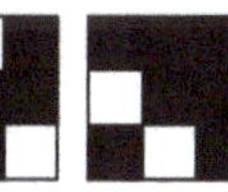   

(A) (B) (C) (D) (E)

The counterclockwise rotation gives us:

The effect of counterclockwise rotation on the five options is shown below.

    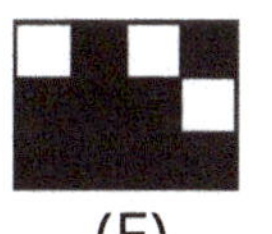

(A) (B) (C) (D) (E)

Figure (D) is the match.

10 **(D) 9 g**
From the first scale, we know that 7 pencils weigh the same as 2 pencils plus 30 grams (g). Remove two pencils from each side of the scale to see that 5 pencils weigh 30 grams, so 1 pencil weighs 6 grams. The second scale shows a pencil plus a pen weighing 15 grams, so the pen weighs 15 grams – 6 grams = 9 grams.

---

11 **(B) 5:05**
The four clocks show the times 4:45, 5:05, 5:25, and 5:40. 5:05 is the real time since one clock (the fast one) shows 5:25 and another clock (the slow one) shows 4:45. The clock showing 5:40 is broken. There are no other options that have two clocks 20 minutes ahead and 20 minutes behind of the indicated time.

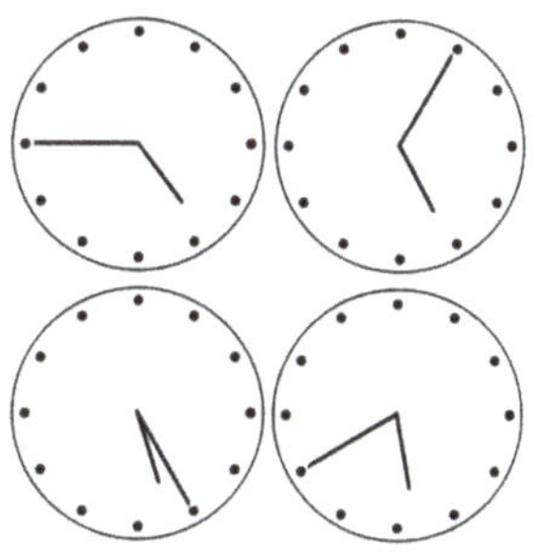

---

12 **(B) more than half of all the fruit.**
Guests ate half of the apples and less than half of the oranges, so they ate less than half of all the fruit. Clearly, more than half of all the fruit remained in the basket.

---

13 **(E) 60,000**
Angie divided the original number by 10 and got 600, so the original number was 6,000. Angie should have multiplied the original number by 10 and the correct result should have been 6,000 × 10 = 60,000.

14 (B) 10

Look at the list 1, 2, 3, ..., 23, 24, 25, ..., 43, 44 (the last missing number). The first 24 numbers are still found in the book, so there are 44 − 24 = 20 missing numbers, which means that there were 20 pages missing (a page is one side of a sheet). Since a sheet has two page numbers, there are 10 sheets missing.

---

15 (E) Friday

Every seventh day after Eva's birthday is a Tuesday, so 49 days (7 weeks) after her birthday it will be Tuesday again. Three days later, 52 days after Eva's birthday, it will be Friday.

---

16 (C) 7

The sum of the first row plus the sum of the second row is the sum of the entire table, so the sum of the entire table is 11. The sum of the first column plus the sum of the second column is also the sum of the entire table. The sum of the second column then is 11 – 4 = 7.

5 Point Solutions

17 (E)

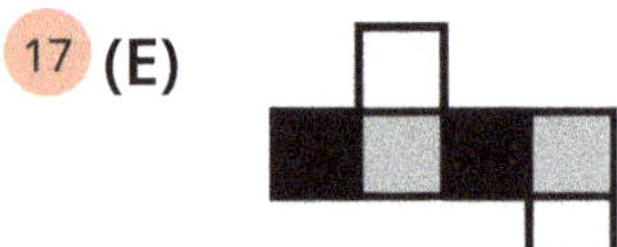

When folding each of the figures shown, the top square and the bottom square will touch all four squares in the middle row, so the top square and the bottom square must be the same color. This excludes all figures except (E).

When folding figure (E), the black squares will be opposite, the gray squares will be opposite, and the white squares will be opposite as well.

18 (C) 64

Label the squares as shown in the picture. The bottom side of square A and the bottom side of square B make up the top side of square C, so 16 plus the length of the bottom side of square B equals 40. So, the length of the bottom side of square B is 40 – 16 = 24. Since B is a square, all sides of square B have lengths equal to 24. The length of the left side of square D is equal to the length of the right side of square B plus the length of the right side of square C. So, the length of the left side of square D equals 24 + 40 = 64.

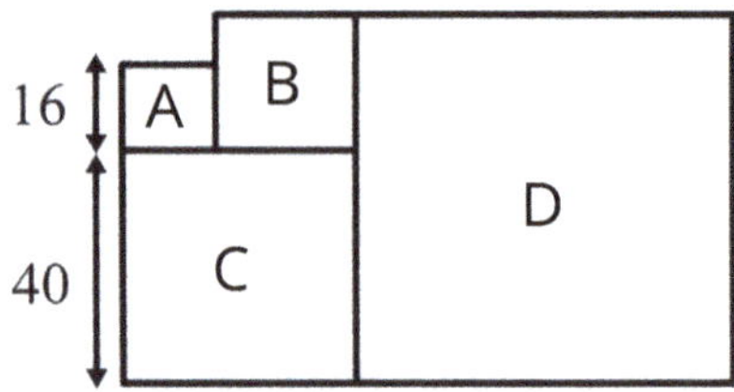

---

19 (C) 23

Draw a circle, labeled Maria, and draw 16 circles to its right. By using the last clue, label the 8th circle to Maria's right as Matt.

Count the 8 students already to the left of Matt and add 6 more to get 14 to his left.

Count the students in the picture. There are 23 students in this class.

20 **(A) 0**

For a 10-digit number to have its digits add up to 9, at least one digit must be 0, because if all digits were at least 1, the sum would be at least 10. Thus, the product of all these digits is 0.

---

21 **(A) 62**

Think of the big cube as a stack of five layers. Each layer contains 5 × 5 = 25 cubes.

The top layer has 13 black and 12 white cubes.

The second layer has 12 black and 13 white cubes (even though you cannot see all the cubes, you know that under each cube in the top layer is a cube of opposite color).

The third and fifth layers are the same as the top, and the fourth layer is the same as the second one. Thus, there are 12 + 13 + 12 + 13 + 12 = 62 white cubes.

---

22 **(B) $375**

Peter paid 3 dollars out of 8 dollars, so the boys split the prize into eight equal parts and Peter got 3 of those equal parts. Each equal part is $125, because 1000 ÷ 8 =125, so Peter received $125 × 3 = $375.

23 **(E) 3**

Focus on the game in which Daniel's team had the one goal scored against them. It could have ended as 0-1, 1-1, 2-1, or 3-1.

In the 0-1 case, the other two games could have ended as 0-0 and 3-0 or 1-0 and 2-0.

In the 1-1 case, the other two games could have ended as 0-0 and 2-0 or 1-0 and 1-0.

In the 2-1 case, the other two games would have ended as 0-0 and 1-0.

In the 3-1 case, the other two games would have ended as 0-0 and 0-0.

Here is the list of all possible results of three soccer games:

0-1, 0-0, 3-0 which is 1 win, 1 tie, 1 loss (4 points);

0-1, 1-0, 2-0 which is 2 wins, 0 ties, 1 loss (6 points);

1-1, 0-0, 2-0 which is 1 win, 2 ties, 0 losses (5 points);

1-1, 1-0, 1-0 which is 2 wins, 1 tie, 0 losses (7 points);

2-1, 0-0, 1-0 which is 2 wins, 1 tie, 0 losses (7 points);

3-1, 0-0, 0-0 which is 1 win, 2 ties, 0 losses (5 points).

Daniel's team could not have earned 3 points.

24 **(E) *M* and *S***

•	*A*	*B*	*C*	7
D	*J*	*K*	*L*	56
E	*M*	36	8	*N*
F	*T*	27	6	*P*
6	18	*R*	*S*	42

•	3	*B*	*C*	7
8	24	*K*	*L*	56
E	*M*	36	8	*N*
F	*T*	27	6	*P*
6	18	*R*	*S*	42

•	3	9	2	7
8	24	*K*	*L*	56
4	*M*	36	8	*N*
3	*T*	27	6	*P*
6	18	*R*	*S*	42

•	3	9	2	7
8	24	72	16	56
4	12	36	8	28
3	9	27	6	21
6	18	54	12	42

To make gray squares easier to reference, the unknown gray numbers have been assigned letters *A* through *F*.

Since each number in the grid is the product of the corresponding gray numbers, $7 \times D = 56$, so *D* is 8. Also, $6 \times A = 18$, so *A* is 3. $A \times D = J$, so $J = 24$.

$C \times E = 8$ and $C \times F = 6$. Only two integers divide both 8 and 6, which are 2 and 1. So *C* equals 2 or 1. If *C* were 1, then *F* would be 6 but then there is no integer solution for $27 = F \times B = 6 \times B$. Therefore, *C* must be 2. Consequently, *F* is 3 and *E* is 4. $F \times B = 27$, so *B* is 9.

Multiplying the numbers in gray allows us fill in the rest of the chart.

12 is the only number that occurs twice, and it is found in the cells labeled *M* and *S*.

2006

3 Point Solutions

1 **(C) 9:00**
It takes 6 hours to go up and back from Mount Giewont (3 hours + 30 minutes + 2 hours 30 minutes). The students need to leave at 9:00 a.m. to get back for lunch at 3 p.m.

2 **(B) 2006**
Remember to do all the multiplication before addition. Any number times 0 is 0.
$2 \times 0 \times 0 \times 6 + 2006 = 0 + 2006 = 2006$.

3 **(D) 7**
There are 9 cubes in the top layer of the first figure. Since there are still 2 cubes left in the top layer of the second figure, 7 cubes have been removed.

4 **(A) Tuesday**
Since tomorrow will be Thursday, today is Wednesday. Katie's birthday was yesterday, which was Tuesday.

5 **(D) 5**
At the beginning John had 10 darts and at the end he had 20, so he gained $20 - 10 = 10$ darts. To gain 10 darts, he had to hit the bullseye $10 \div 2 = 5$ times.

6 **(E) e**

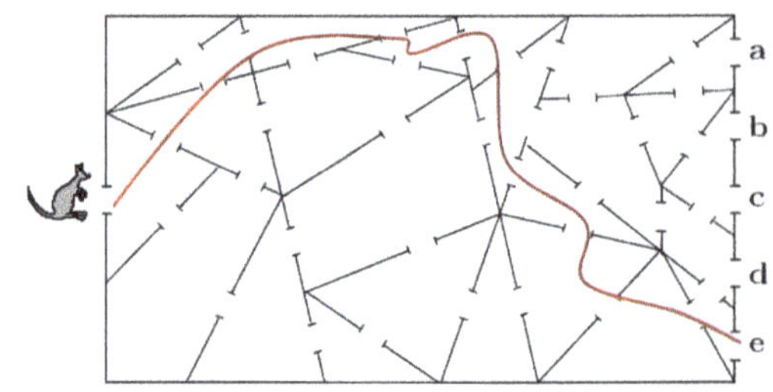

7 **(B) 16**

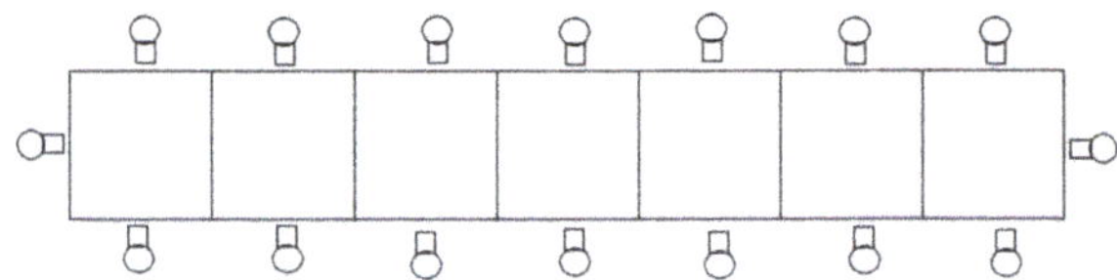

8 **(B) $4**
$3 = $1 + $2
$6 = $5 + $1
$7 = $5 + $2
$8 = $5 + $2 + $1.

Only $4 cannot be expressed as a sum of $1, $2, $5.

4 Point Solutions

9 **(C) 17**

The odd-numbered houses are 1, 3, 5, 7, 9, 11, 13, 15, 17, and 19; that is 10 houses.

The even-numbered houses are 2, 4, 6, 8, 10, 12, 14; that is 7 houses.

10 + 7 = 17.

10 **(A)**

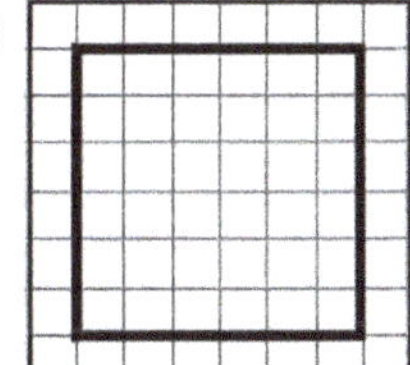

The figure needs to be cut out from a rectangle that is at least 6 squares tall and 6 squares across. Only (A) shows a figure large enough both in height and width.

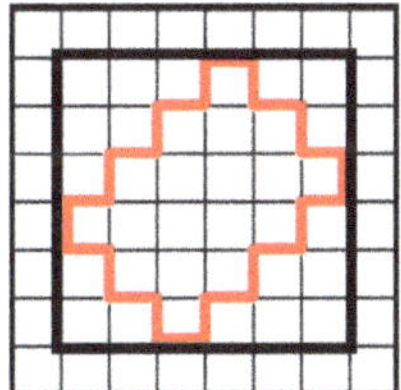

---

11 **(A) 90**

90 = 20 + 10 + 30 + 20 + 10

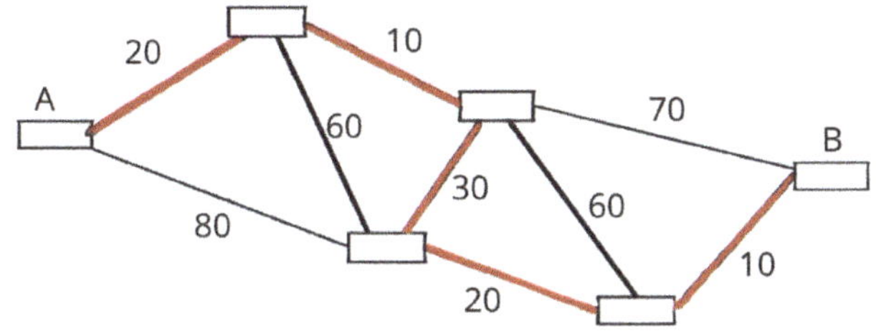

12 **(D) 2309415687**

The digits on the left side of the number have a greater value based on place value, so we need to start with the smallest digit possible. So, we have to start with the card with number 2. Afterwards, always chose a card from those that remain with the smallest first digit you can find. The remaining cards should be arranged as follow: 309, 41, 5, 68 and 7.

---

13 **(C) 3 and 1**

The six weights together weigh 21 pounds. The weights in the first and second box together weigh 17 pounds, so the weights in the third box must weigh 4 pounds. The only option is to have the 1-pound and 3-pound weights in the third box. The first box contains the 4-pound and 5-pound weights, and the second box contains the 2-pound and 6-pound weights.

---

14 **(E) All are equal.**

All the routes are equal. All have the length of 8 diagonals of a unit square.

---

15 **(A) 46**

8 = 1 × 6 + 2
18 = 3 × 6 + 0
28 = 4 × 6 + 4
38 = 6 × 6 + 2
48 = 8 × 6 + 0
58 = 9 × 6 + 4

The only numbers with the remainder 2 are 8 and 38. These are the numbers on the petals Mary picked off. Their sum is 38 + 8 = 46.

16 (C)

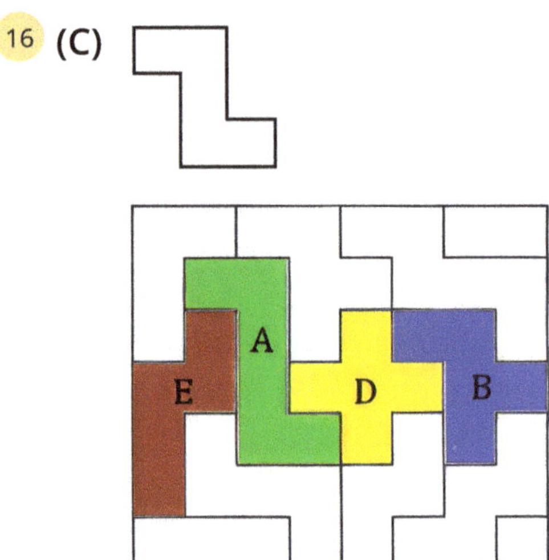

5 Point Solutions

17 **(B) 6 feet**

The distance between Hanna and Dana is the same as the distance between Dana and Lena, so the distance between Lena and Bennie is the same as the distance between Dana and Lena.

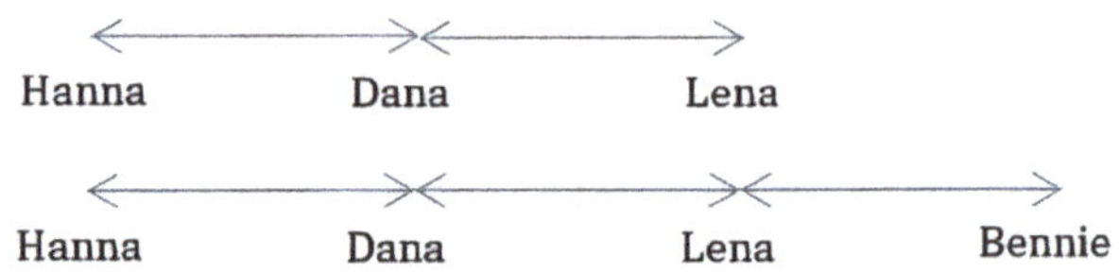

Hence, Lena sits exactly in the middle between Dana and Bennie, so the distances between Hanna and Dana, Dana and Lena, and Lena and Bennie are the same. The distance between Dana and Bennie is twice the distance between Dana and Lena, so the distance between Dana and Lena is half the distance between Dana and Bennie. The distance between Dana and Bennie is 4 feet, so the distance between Dana and Lena is 2 feet. The distance between Hanna and Bennie is 3 times the distance between Dana and Lena which is 6 feet, so Hanna is sitting 6 feet from Bennie.

18 **(D) 26**

To build the bottom level of a 4-story house from the 3-story house shown to the left, we have to add 3 cards in flat positions (one of them is shown in orange) and below these 3 cards we have to add 4 pairs of cards such as used in the one-story house. Together, 11 new cards are added to the 15 cards of the 3-story house, so Johnny needs 26 cards to build a 4-story house.

19 **(D) 36**

The surface of the structure consists of top faces of six cubes, left faces of six cubes, and front faces of six cubes. The same is true for bottom, right, and back faces, so Roman painted $6 \times 6 = 36$ faces of the cubes.

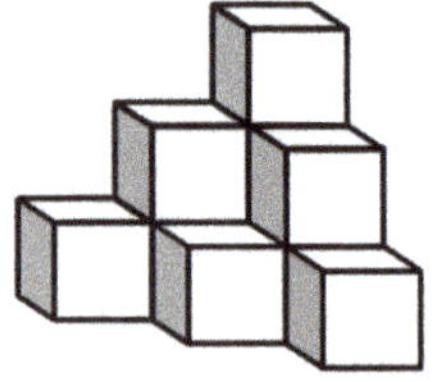

20 **(E) Ann and Olga**

Kate and Elena live on the same floor since Olga lives on the other floor. Irena and Kate live on the same floor since Ann lives on the other floor. Thus Kate, Elena, and Irena live on the same floor, which must be the second floor. Therefore, Ann and Olga live on the first floor.

21 (B) 2001

If we add or subtract two odd numbers, the result is an even number. By adding or subtracting only even numbers (doesn't matter how many), the result is always an even number. In our expression we only have two odd numbers, so for any distribution of "+" and "–" the result is always an even number which makes 2001 an impossible result. The rest of the results can be made as follows:

1998 = 2002 + 2003 + 2004 – 2005 – 2006
2002 = 2002 + 2003 – 2004 – 2005 + 2006
2004 = 2002 – 2003 + 2004 – 2005 + 2006
2006 = 2002 – 2003 – 2004 + 2005 + 2006.

---

22 (E) Thursday

March has 31 days. If March has 5 Mondays, then the first Monday of the month cannot happen on March 4th or later. Otherwise, there would not be a fifth Monday in March since 4 + (7 + 7 + 7 + 7) = 32.

There are 3 options for the first Monday of the month, March 1st, March 2nd, or March 3rd.

Monday	1	8	15	22	29
Tuesday	2	9	16	23	30
Wednesday	3	10	17	24	31

Sunday	1	8	15	22	29
Monday	2	9	16	23	30
Tuesday	3	10	17	24	31

Saturday	1	8	15	22	29
Sunday	2	9	16	23	30
Monday	3	10	17	24	31

The tables show other days of the week that also appear 5 times. Neither Thursday nor Friday appear 5 times in any of these cases, so Thursday is the answer, since Friday is not among the options listed.

23 (C) 4

We have four starting options as shown below. Each of the starting options extends to only one particular solution as shown.

1	2	
2		

1	3	
2		

1	2	
3		

1	3	
3		

→

1	2	3
2	3	1
3	1	2

1	3	2
2	1	3
3	2	1

1	2	3
3	1	2
2	3	1

1	3	2
3	2	1
2	1	3

---

24 (B) 20

To balance the right part, the trapezoid shape must weigh 60 ounces, so the weights of the whole right part weigh 120 ounces.

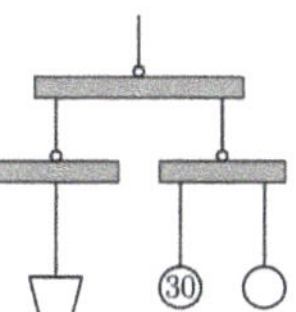

To balance the left part, the heart-shaped weight weighs twice as many ounces as the square shape, so the weight of the whole left part (two hearts and two squares) is equal to the weight of 6 square shape weights (the two squares shown plus two more for each of the hearts). To balance 120 ounces of the right part the square shape must weigh 20 ounces since 6 × 20 ounces is 120 ounces.

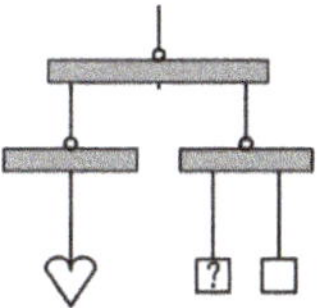

A complete graphic solution is shown below.

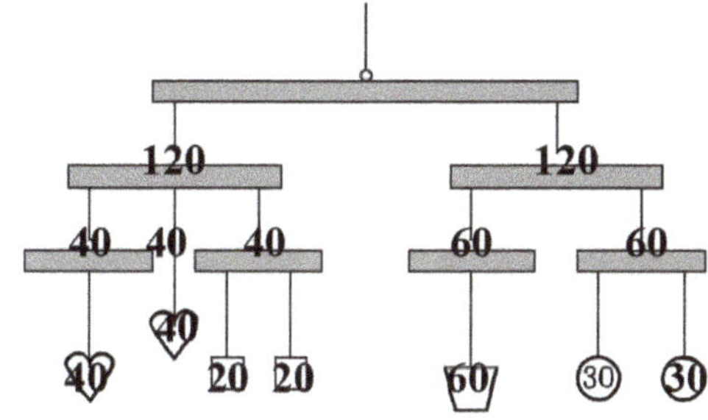

2008

3 Point Solutions

1 **(C) 21**
There are 7 days in a week and Ann eats 3 pieces of candy per day, so she eats $7 \times 3 = 21$ pieces of candy in a week.

2 **(D) $10**
An adult ticket costs $4 and a child ticket costs $4 – $1 = $3, so one adult and two child tickets cost $4 + $3 + $3 = $10.

Father's ticket   Children's tickets

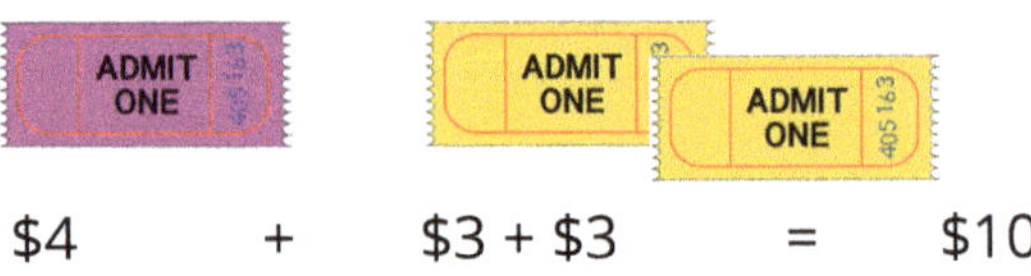

$4 + $3 + $3 = $10

3 **(B) pink roses**

We know that the flowers Luke's aunt and sisters received were of the same color, so these three women received yellow flowers. We also know that Luke's grandmother did not receive roses, so she received red carnations. The bouquet that Luke's mother received were pink roses.

4 **(B) 17**
$37 - 10 = 27$, so Adelaide would be left with 27 CDs after giving 10 of them to Mary. At that moment Mary would have 27 CDs and 10 of them are from Adelaide, so she had 17 CDs of her own.

5 **(D) 8**
1 line passing through the point divides the rectangle into 2 pieces, 2 lines divide the rectangle into 4 pieces, 3 lines divide the rectangle into 6 pieces, and 4 lines divide the rectangle into 8 pieces. Each time you add a new line passing through the point, it increases the number of pieces by 2.

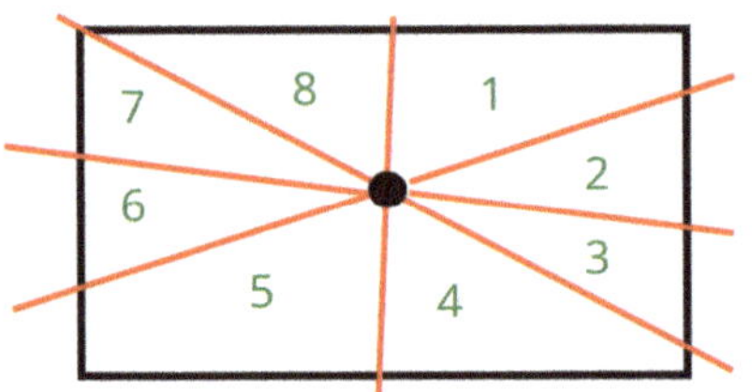

6 **(A) 9:30**
6 hours before 4:00 is 10:00 (4 hours back is 12:00, and another 2 hours back makes it 10:00). Going back 30 minutes from 10:00 gives 9:30.

7 **(E)**

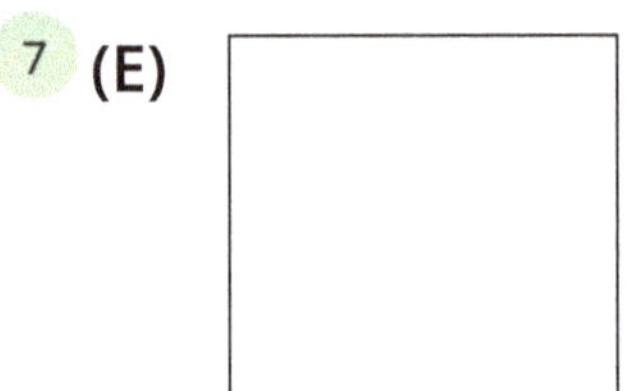

Charlie can get each of the first four figures, but he can't get the last one. If he places two equilateral triangles in any way he wants, his figure will have at least one angle equal to 60°.

8 **(A) 57**
Before the storm, the side of the roof shown in the picture was made of $7 \times 10 = 70$ tiles and the hole was covered by (counting by rows) $3 + 4 + 4 + 2 = 13$ tiles. The number of tiles left is $70 - 13 = 57$ tiles.

4 Point Solutions

9 **(E) 1988**

The seventh Kangaroo took place 10 years before the seventeenth Kangaroo, so the seventh Kangaroo happened in 1998. Maggie was 10 in 1998, so she was born in 1988.

Math Kangaroo	1992	1993	1994	1995	1996	1997	1998	1999	2000	2001	2002	2003	2004	2005	2006	2007	2008
	1st	2nd	3rd	4th	5th	6th	7th	8th	9th	10th	11th	12th	13th	14th	15th	16th	17th

---

10 **(D) Jim, Greg, Michael**

The three boys can perform their operations in the following orders:

Greg, Jim, Michael
$[(3 \times 3) + 2] - 1 = (9 + 2) - 1 = 11 - 1 = 10$

Greg, Michael, Jim
$[(3 \times 3) - 1] + 2 = (9 - 1) + 2 = 8 + 2 = 10$

Jim, Greg, Michael
$[(3 + 2) \times 3] - 1 = (5 \times 3) - 1 = 15 - 1 = 14$

Jim, Michael, Greg
$[(3 + 2) - 1] \times 3 = (5 - 1) \times 3 = 4 \times 3 = 12$

Michael, Greg, Jim
$[(3 - 1) \times 3] + 2 = (2 \times 3) + 2 = 6 + 2 = 8$

Michael, Jim, Greg
$[(3 - 1) + 2] \times 3 = (2 + 2) \times 3 = 4 \times 3 = 12$

There is only one way to end up with 14, so the boys should perform their operation as follows: First Jim adds 2, then Greg multiplies by 3, and finally Michael subtracts 1.

---

11 **(E) Tania**

Grace is taller than Ann and taller than Irena. Irena is taller than Kate, so Grace is taller than Ann, Irena, and Kate. Since Grace is shorter than Tania, Tania is the tallest.

12 **(D)**

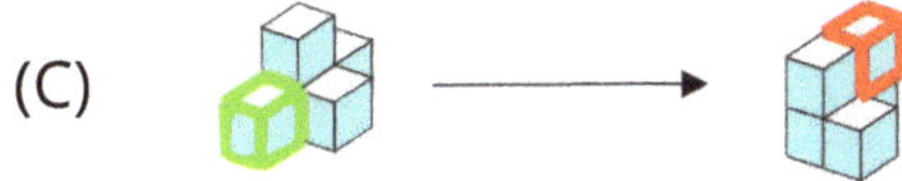

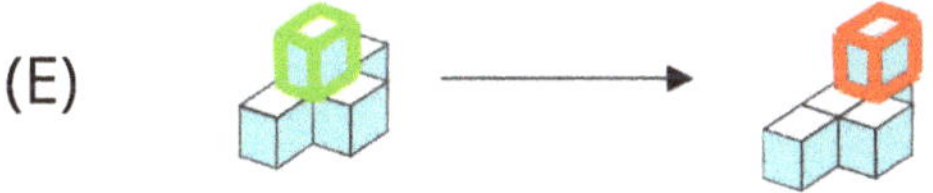

---

13 **(E) 23**

If we take any two different numbers, one is a smaller number and the other is a larger number. In this problem, the sum of the two numbers is 45. The double of the smaller number is less than 45. The double of the larger number is more than 45. Thus, the smaller number is always less than half of 45 and is at most 22, which means it has to be less than 23.

(The larger number is always more than half of 45, so it is at least 23.)

All possible options of two different numbers with the sum of 45 are shown:

1 + 44	12 + 33
2 + 43	13 + 32
3 + 42	14 + 31
4 + 41	15 + 30
5 + 40	16 + 29
6 + 39	17 + 28
7 + 38	18 + 27
8 + 37	19 + 26
9 + 36	20 + 25
10 + 35	21 + 24
11 + 34	22 + 23

14 **(C) 3**
5 three-person rooms can hold 5 × 3 = 15 people. 21 − 15 = 6, and the hotel must have 6 ÷ 2 = 3 two-person rooms to hold the 6 people.

---

15 **(B) 29 minutes 3 seconds**

6 min 25 sec
12 min 25 sec
+ 10 min 13 sec

28 min 63 sec = 29 minutes 3 seconds

---

16 **(B) 6**

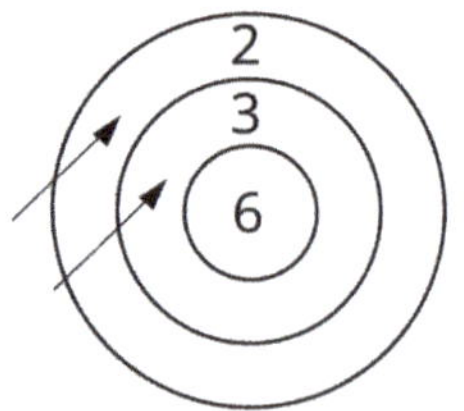

The possible scores are 4, 5, 6, 8, 9, and 12.

2 + 2 = 4
2 + 3 = 3 + 2 = 5
2 + 6 = 6 + 2 = 8
3 + 3 = 6
3 + 6 = 6 + 3 = 9
6 + 6 = 12

There are 6 different possible scores.

5 Point Solutions

17 **(C) 8**

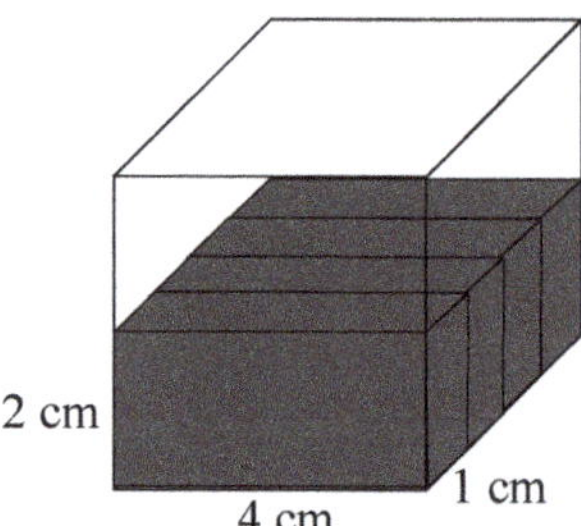

On the bottom, we can place 4 blocks with the dimensions 1 cm × 2 cm × 4 cm.

On the top, we can place a second layer of 4 blocks with the dimensions 1 cm × 2 cm × 4 cm.

18 **(A) 92 kilograms**
We can work backwards.

2004				2005				2006				2007				2008	
WINTER	SPRING	SUMMER	FALL	WINTER	SPRING	SUMMER	FALL	WINTER	SPRING	SUMMER	FALL	WINTER	SPRING	SUMMER	FALL	WINTER	SPRING
			92 kg		97 kg		93 kg		98 kg		94 kg		99 kg		95 kg		100 kg

-5 kg +4 kg -5 kg +4 kg -5 kg +4 kg -5 kg

---

19 **(D) 16 m**
The perimeter is the distance around a figure. For a square, it is four times the length of one side, because there a square has four sides of equal length. For a rectangle, the perimeter is equal to two times the length of the shorter side added to two times the length of the longer side, because opposite sides are equal.

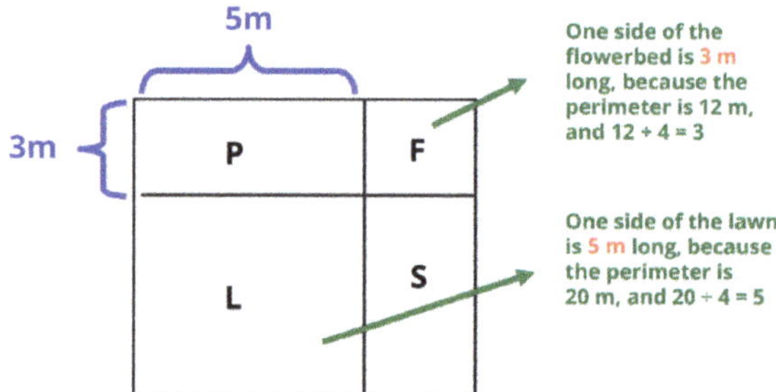

The perimeter of the pool is 16 m since 2 × 3 + 2 × 5 = 16.

---

20 **(E) 7**
Diane has twice as many brothers as sisters. One of her brothers is Henry. If Diane had only one sister, then she would have to have two brothers. In this case, Henry would have only one brother but two sisters, which can't be correct because Henry has as many brothers as sisters.

Suppose that Diane has 2 sisters. Then, she would have to have 4 brothers. If this is the case, then Henry has 3 sisters and 3 brothers, which make the numbers of this brothers and sisters the same. So, there are 3 girls and 4 boys in the family, which means that there are 3 + 4 = 7 children.

21 **(E) 36**

12, 13, 14, 15, 16, 17, 18, 19,	→	*8*
23, 24, 25, 26, 27, 28, 29,	→	*7*
34, 35, 36, 37, 38, 39,	→	*6*
45, 46, 47, 48, 49,	→	*5*
56, 57, 58, 59,	→	*4*
67, 68, 69,	→	*3*
78, 79,	→	*2*
89, +	→	*1*
		36

22 **(B) 20**

The green segment below represents Li's age and the orange segment represents the period of 6 years. The whole segment represents Li's age in 6 years.

Li's age in 6 years

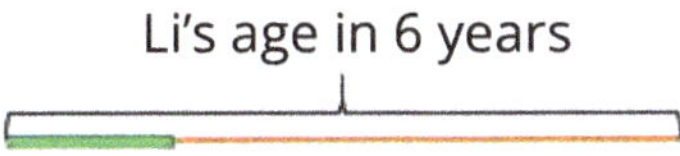

If we double the whole segment, we get Mary's age in 6 years, so Mary's age now is represented by the green-orange-green interval.

Mary's age in 6 years

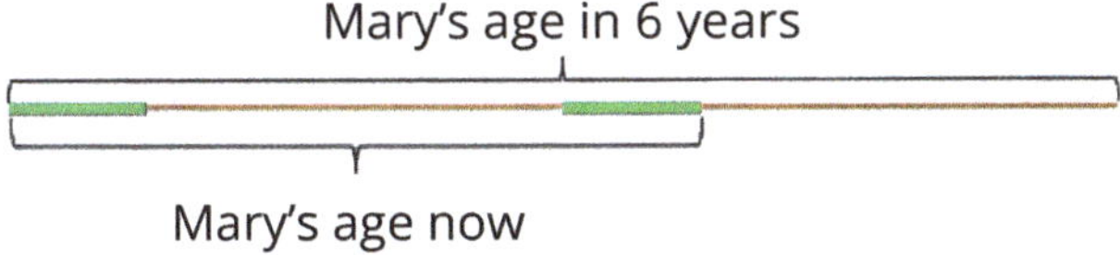

Mary's age now

Mary's age is also represented by 5 green segments since Mary is five times as old as Li.

Two representations of Mary's age

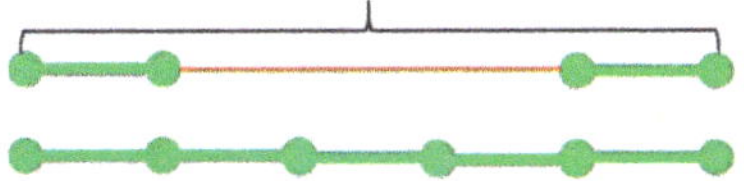

Hence, the orange segment is equivalent to 3 green segments.

The orange segment represents the period of 6 years, so Li's age is 2, Mary's age is $5 \times 2 = 10$ and in ten years Mary will be $10 + 10 = 20$ years old.

23 **(C) 8**

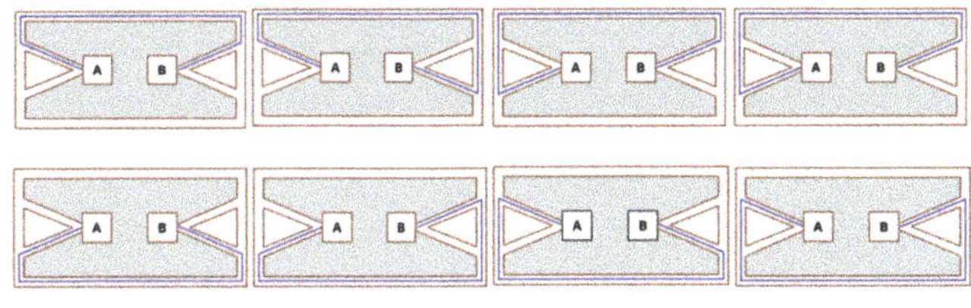

24 **(D) 25**

Let's reverse the stages of pouring the water back and forth.

At each stage the total amount of water in the two containers is 40 liters.

At the end, the total of 40 liters of water is split evenly between two containers. Each container has 20 liters of water and in the first container, C-1, the amount of water was doubled compared to the previous stage, so 10 liters of water were in the first container before this.

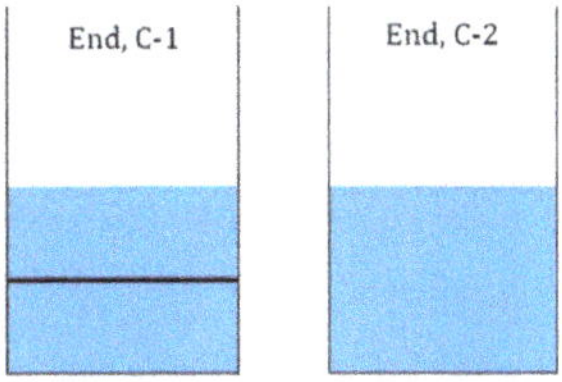

Before the middle stage 5 liters of water were poured from the first container to the second container.

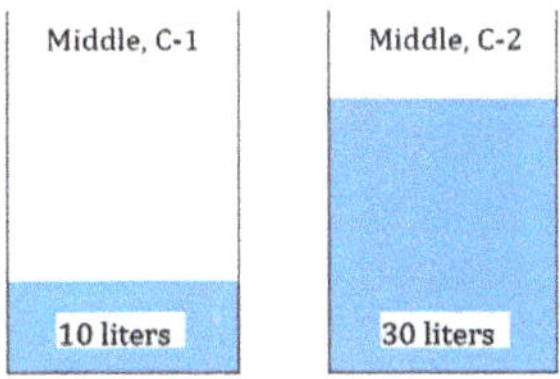

At the beginning, there were 15 liters of water in the first container, and 25 liters of water in the second container.

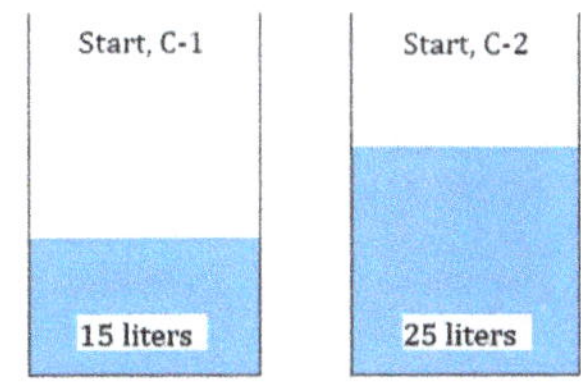

2010

3 Point Solutions

1 **(D) 2 − 0 + 1 + 0**

2 + 0 − 1 + 0 = 2 − 1 = 1
2 − 0 − 1 + 0 = 2 − 1 = 1
2 + 0 − 1 − 0 = 2 − 1 = 1
2 − 0 + 1 + 0 = 2 + 1 = 3
2 − 0 − 1 − 0 = 2 − 1 = 1

---

2 **(C) 12:10**
Since the lesson lasts 40 minutes, half-way through is 20 minutes into the lesson. 20 minutes after 11:50 is 12:10.

---

3 **(C) Daniel's**
The person whose steps are the longest needs to make the least number of steps to cover a given distance, so Daniel's steps were the longest.

---

4 **(C) 31**
The first day, Adam solved 1 problem. The second day, he solved 2 × 1 = 2 problems.

The third day, he solved 2 × 2 = 4 problems. The fourth day, he solved 2 × 4 = 8 problems.

The fifth day, he solved 2 × 8 = 16 problems. Altogether, he solved 1 + 2 + 4 + 8 + 16 = 31 problems.

---

5 **(D)** 

Notice that in picture (D) the lines run across the small squares and not from corner to corner, like the line does in the picture of the piece of cardboard. All the other pictures can be easily made from the 4 pieces of cardboard.

6 **(A) \$3**
Buying the items separately would cost \$4 + \$9 + \$5 = \$18. So, by buying the set mother would save \$18 − \$15 = \$3.

---

7 **(C) 54**
Work backwards. Before Eva bought the 16 pairs of shoes, she did not have shoes on 7 + 16 = 23 pairs of feet. This means that she had shoes on 50 − 23 = 27 pairs of feet, so she had 2 × 27 = 54 feet with shoes on before she bought the 16 pairs of shoes.

---

8 **(B) 2**

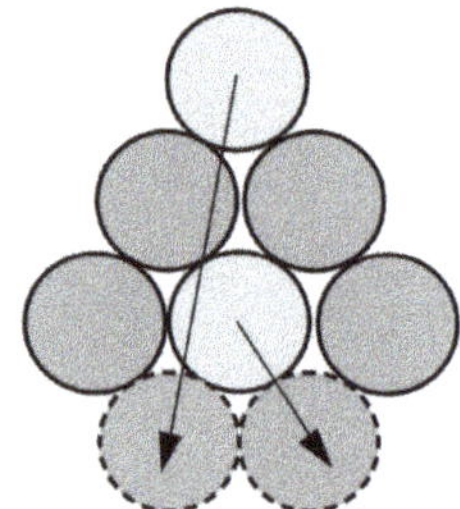

4 Point Solutions

9 **(E) the number of seconds in one week.**

There are 60 seconds in one minute, 60 minutes in one hour, 24 hours in one day, and 7 days in one week, so 60 × 60 × 24 × 7 is the number of seconds in one week.

For the other options we have the following expressions:

(A) 60 × 24 × 7 × 7
(B) 60 × 60 × 7
(C) 60 × 24 × 7 × 24
(D) 24 × 60

10 **(B) Adam, Alex, Tom, Luke**

From the statement, "Tom ate more ice cream than Luke but less than Alex," the order of these three boys from the one who ate the most to the one who ate the least is Alex, Tom, Luke. Since Adam ate more ice cream than Alex, Adam ate the most ice cream.

---

11 **(B) 14**

When Matthew was on the 8th floor, he was already half way through his 12-floor climb, so he still had 6 more floors to go. Hence, Clara lives on 8 + 6 = 14th floor.

---

12 **(D) 20**

Along each vertical edge there are 3 cubes with exactly two sides painted, so along the four vertical edges there are 4 × 3 = 12 such cubes. Along each horizontal bottom edge there are 2 cubes with exactly two sides painted, so along the four bottom edges there are 4 × 2 = 8 such cubes. There are no additional cubes like these along the top horizontal edges.

Altogether, there are 12 + 8 = 20 cubes with exactly two sides painted.

---

13 **(D) $200**

The perimeter of the fence to be painted is 2 × 20 meters (two longer sides) + 10 meters (one shorter side) which is 50 meters. Using one can of paint Grandpa can paint 10 meters of the fence since it takes one-half of a can to paint 5 meters of the fence. Using five cans of paint Grandpa can paint the whole fence since 50 meters = 5 × 10 meters. One can of paint costs $40, so Grandpa will pay 5 × $40 = $200 for the paint he needs.

14 **(C)**

			69	
	72			

The ones digits of the numbers in any row are:

1 2 3 4 5 or 6 7 8 9 0

This excludes (A), (D), and (E) since (A) has 3 as the ones digit in the second column, (D) has 1 and 6 as the ones digits in the second column, and (E) has 7 as the ones digit in the third column. (B) is excluded since the numbers in the table are increasing from one row to the next.

(C) represents the 14th and 15th rows of the table.

66	67	68	69	70
71	72	73	74	75

---

15 **(D) 20**

At each corner of a little square there are two edges labeled with numbers. One of the numbers is odd and the other number is even. In the clockwise direction the order is always the odd number first and then the even number next. Anne always cut along the edges labeled with even numbers, so the sum is 2 + 4 + 6 + 8 = 20.

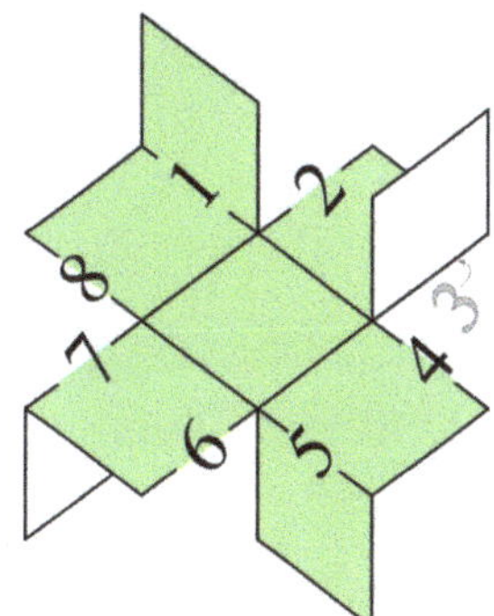

16 **(A) 99**

Notice that each of the numbers shown in the bottom row of the table is 10 more than the number above it. Since there are 10 numbers shown in the bottom row, the difference between the two rows, not counting the last column, is 10 × 10 = 100. So, to make the sums in both rows equal, the last number must be 100 less than the number above it. The value of * is 199 − 100 = 99.

5 Point Solutions

17 **(E)**

There is a cut at every place that was a corner when the piece of paper was folded, which means that the four corners of the large square are clipped and there is a hole in the middle.

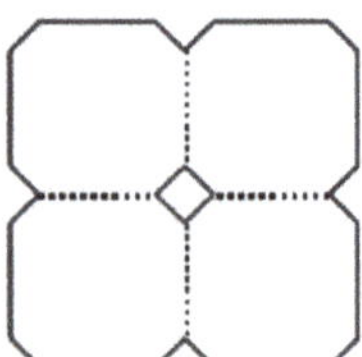

18 **(C) 2003**

Since none of the children were right, and Anna's guess of 2010 differs from 1998 by 12 and from 2015 by 5, her guess must differ from the right number by 7 (otherwise, the right answer would have been 1998 or 2015, and we know it was not the case). So, the number of the books was either 2010 − 7 = 2003 or 2010 + 7 = 2017. It cannot be 2017 since 2017 − 1998 = 19, which is neither 12 nor 5. The other option is 2003. 2003 − 1998 = 5 and 2015 − 2003 = 12, so there are 2003 books in the library.

19 **(D) 22**

Tom counted 12 − 3 = 9 chairs before he got to where Adam started counting. Adam counted 18 − 5 = 13 chairs before he got to where Tom started counting. There were 9 + 13 = 22 chairs around the table.

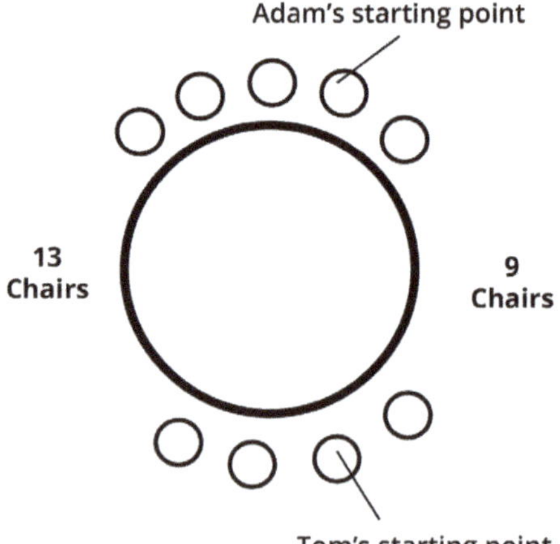

20 **(D) 16 mm**

The two-link overlap has the width of 2 × 0.5 mm = 1 mm. The first link has the length of 4 mm and each additional link will only increase the length of the chain by 4 mm – 1 mm = 3 mm.

A chain made up of five links will be 4 mm + 4 × 3 mm = 4 mm + 12 mm = 16 mm long.

21 **(E) 5**

At least one red and one yellow ladybug came to the party. Together this pair of ladybugs had 6 + 10 = 16 spots. The other ladybugs at the party had a total of 42 − 16 = 26 spots.

26 is neither a multiple of 6 nor a multiple of 10, so among the other ladybugs there must be another pair of ladybugs of different colors. Together, the two pairs of ladybugs had 2 × 16 = 32 spots, so 42 − 32 = 10 spots were not counted yet. All these 10 spots belong to one yellow ladybug, so 3 yellow ladybugs and 2 red ladybugs came to the party, which is 5 ladybugs altogether.

22 **(B) 8**

The number of days in any month does not exceed 31. To get the sum of 35, the number of the month needs to be at least 4. Since April is the 4th month and only has 30 days, it cannot be included. Basil's friends had to be born in the 8 different months from May, which is the 5th month, to December, so 8 is greatest number of Basil's friends.

23 **(D) Washington**

Paul is not from Washington, Pittsburgh, or New York, so he must be from Dallas. Micah is not from Dallas and he is not from Washington or Pittsburgh, so Micah is from New York. Jeff is not from Pittsburgh, so he is from Washington.

24 **(E) 96**

Olivia's grandmother is 10 × 6 = 60 years old. Olivia's mother is 60 − 14 − 10 = 36 years old.

Olivia's great-grandmother is 60 + 36 = 96 years old.

2012

3 Point Solutions

1 **(B) 8**
There are 8 different letters in the word MATHEMATICS: M (twice), A (twice), T (twice), H, E, I, C, and S, so Basil will need 8 colors.

---

2 **(D)**

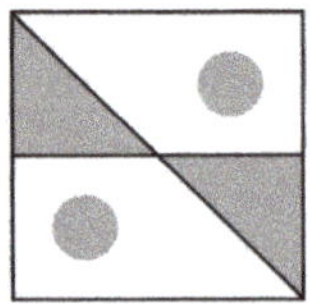

The blue vertical line and the inside horizontal line divide the original square into four corner squares. The upper left square and the lower right square are divided evenly into the white and the gray areas. Each of the other two corner squares show more white than gray area, so in (D) the white area and the gray area are different.

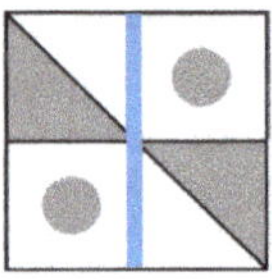

---

3 **(B) 10**
4 pins are needed for 3 towels. The next towel will use one pin from a towel that is already on the line and one new pin at the other end, so 5 pins are needed for 4 towels. Any new towel will use one pin which is already on the line at one end and one new pin at the other end, so the father needs one additional pin for each additional towel at this point. Since he needs to hang up 9 – 3 = 6 more towels, he needs 6 more pins, so he needs 4 + 6 = 10 pins total.

---

4 **(C)**

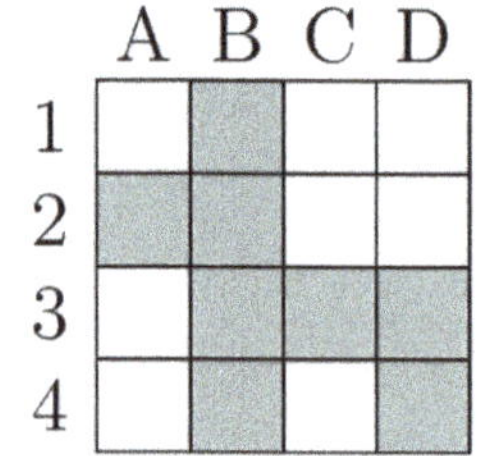

The letter tells us in which column the colored square is, and the number tells us in which row. Among all patterns shown below only (C) has the square B1 colored. The other squares of (C) are A2, B2, B3, C3, D3, B4, and D4 which, together with B1, are exactly the squares colored by Iljo.

---

5 **(A) 3**
One of 13 children playing is the “seeker” and the other 12 children were hiding. 9 children have been found, so 12 – 9 = 3 are still hiding.

---

6 **(E) Mike; he scored 4 points more.**
Mike’s points are 25 + 35 + 7 = 67. Jake’s points are 15 + 45 + 3 = 63.
Mike won; he scored 67 – 63 = 4 points more than Jake.

---

7 **(B) 8**
Each row of the rectangular pattern had 4 gray tiles and 4 striped tiles, so there were 4 × 5 = 20 gray tiles. 12 gray tiles still remain, so 20 – 12 = 8 gray tiles have fallen off.

---

8 **(D) on February 24**
Ducklings which are 20 days old on March 15 had to hatch 20 days before that day. There are 14 days in March before March 15, so we need to count the last 20 – 14 = 6 days in February. These days are 29th, 28th, 27th, 26th, 25th and 24th, so the ducklings hatched from their eggs on February 24.

4 Point Solutions

9 **(E) 4**

The construction of all 4 shapes (one piece green, the other gray) is shown below:

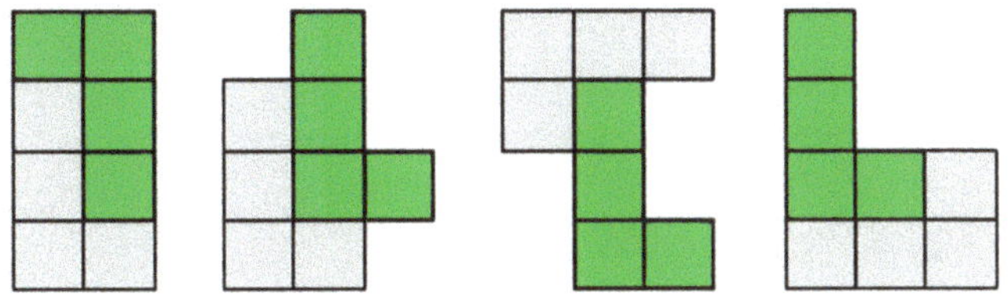

---

10 **(B) 6**

The difference in cost between 3 balloons and 1 balloon is 12 cents, so 2 balloons cost 12 cents. Therefore, one balloon costs 6 cents.

---

11 **(E) 10**

After Grandmother decorated 15 cookies with raisins, there were still 5 plain cookies left. To get the smallest number of cookies decorated with both raisins and nuts, she first has to decorate the 5 plain cookies with nuts. That forces her to decorate 10 more cookies already decorated with raisins, so 10 is the smallest number of cookies decorated both with raisins and nuts.

---

12 **(C) 3**

Let's evaluate the expressions as they appear in the sudoku square.

1		3	
4	3		1
3	4	1	
2	1		

In the 2nd row the missing number is 2. Write it in and look to the right.

1		3	
4	3	2	1
3	4	1	
2	1		

Now in the 3rd column the missing number is 4. Write it in and look at the last row.

1		3	
4	3	2	1
3	4	1	
2	1	4	

The missing number in the lower right corner is 3.

1	2	3	4
4	3	2	1
3	4	1	2
2	1	4	3

The complete solution is shown to the left.

13 **(D) 25**

Nikolay's classmates can be split into groups with 2 girls and 1 boy in each group. Each group consists of 3 of Nikolay's classmates, so the number of all children in this class (excluding Nikolay) is a multiple of 3. Subtract 1 from the possible answers and check which result is a multiple of 3. 30 − 1 = 29, 20 − 1 = 19, 24 − 1 = 23 and 29 − 1 = 28 are not multiples of 3.

Only 25 − 1 = 24 is a multiple of 3 (24 = 8 × 3). 25 can be the number of children in the class.

---

14 **(B) 5**

3 kittens have 3 × 4 = 12 legs, 4 ducklings have 2 × 4 = 8 legs, and 2 baby geese have 2 × 2 = 4 legs. Together these animals have 12 + 8 + 4 = 24 legs, so the lambs have 44 - 24 = 20 legs. Since each lamb has 4 legs, there were 20 ÷ 4 = 5 lambs in the animal school.

---

15 **(D)**

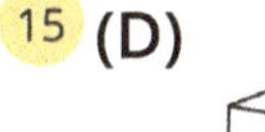

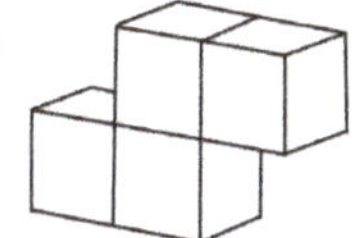

At least one face of each light gray cube is shown. The same is true about each dark gray cube. One black cube is hidden but it is connected to the 3 black cubes shown. The only space for it is under the white cube which is adjacent to the black cube (see the top face of the prism). There are only two spaces available for the two hidden white cubes, one under the other white cube visible on the top face of the prism and the other next to it under the dark gray corner cube at the top and back of the prism. The shape of the white piece is shown in (D).

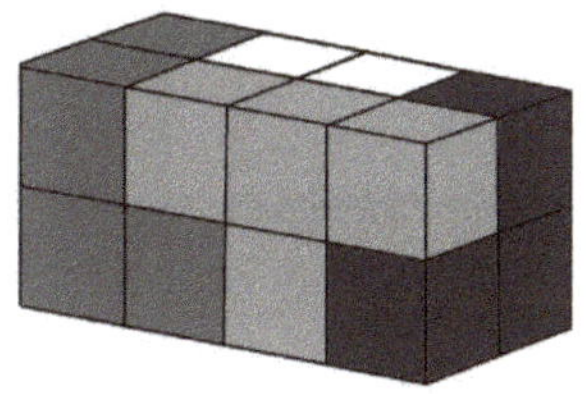

16 **(C) 57**
There were 6 five-branched candlesticks and 15 – 6 = 9 three-branched candlesticks, so 6 × 5 + 9 × 3 = 30 + 27 = 57 candles had to be bought for all the candlesticks.

5 Point Solutions

17 **(D) 12**
The grasshopper can jump up 7 times reaching the 21st step. At this point she can jump down landing on the 17th step. Now, she can jump up reaching the 20th step. The next jump must be down to the 16th step and from there she will make 2 jumps up to reach the 22nd step.

3 + 3 + 3 + 3 + 3 + 3 + 3 – 4 + 3 – 4 + 3 + 3 = 22 and 10 jumps up + 2 jumps down is 12, which is the smallest number of jumps to take a rest on the 22nd step.

The grasshopper can change the order of up and down jumps as long as she does not go below the ground level.

18 **(C) 4**
The number of dots on each tile is shown here by numbers. Adjacent parts of different tiles have the same numbers, so the left missing tile has 2 dots on its left side and the right missing tile has 1 dot on its right side (see both in red).

The sum of the dots already accounted for is: 25 = 1 + 0 + 0 + 5 + 5 + 3 + 3 + 2 + **2** + **1** + 1 + 2.

33 − 25 = 8, and since the two missing tiles have the same number touching, the ? has 4 dots.

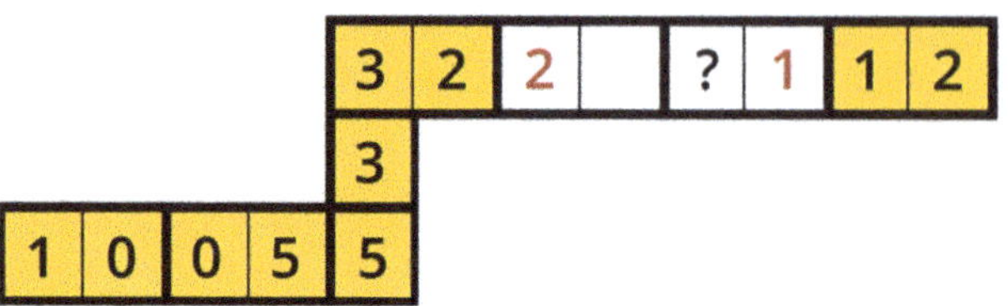

19 **(D) 1173**
To create the largest sum, the numbers will have the largest two digits (6 and 5) in the hundreds place, the next largest two digits (4 and 3) in the tens place, and the smallest two digits (2 and 1) in the ones place. There are a few ways to do this (for example 632 and 541), all of which give the same sum of 1173.

20 **(B) 4**
If Laura stands to the right of Kate, then Iggy will stand to the right of Laura. Val has two ways to be in the picture, at either end of the group.

If Laura stands to the left of Kate, then Iggy will stand to the left of Laura. Again, Val has two ways to be in the picture, at either end of the group.

This group can pose for the picture in 4 different ways: VKLI, KLIV, VILK, or ILKV.

21 **(E)**

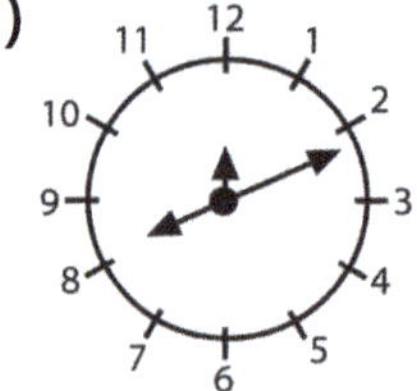

12:55:30 is almost one o'clock, so the medium hand is the hour hand. At 55 minutes past the hour the minute hand will be pointing to the 11, so the longest hand is the minute hand.

The shortest hand shows the seconds.

At 8:11:00 the hand showing seconds points up directly toward 12, so the above options (A), (B), and (D) are eliminated right away. At 8:11:00 the minute hand must be a little past the 10 minutes mark represented by 2 on the clock. This also eliminates (C). (E) shows the hour just a bit past eight o'clock, so that is the clock that shows 8:11:00.

22 **(D) 7**

Let's reverse the process. Since Michael multiplied by 4 in the last operation to get the final answer of 2012, we need to divide 2012 by 4, which results in 503. Since he added 3 in the previous step, we need to subtract 3 from 503, which equals 500. Before that, he multiplied his previous result by 10, so we will divide 500 by 10, which gives us 50. He added 1, so we will subtract 1 from 50, which results in 49. The number which Michael chose at the beginning needs to result in 49 when multiplied by itself. That number is 7, because 7 × 7 = 49.

---

23 **(E) 12 mm**

From the 84 × 192 rectangle we can cut off two 84 × 84 squares since 84 + 84 = 168 is less than 192. What remains is a 84 × 24 rectangle since 192 − 168 = 24.

From the 84 × 24 rectangle we can cut off three 24 × 24 squares since 24 + 24 + 24 = 72 is less than 84. What remains is a 12 × 24 rectangle since 84 − 72 = 12. The 12 × 24 rectangle can be split into two 12 × 12 squares, and now there are no more non-squares to cut, so 12 mm is the length of the side of the smallest square.

In the picture, the two 84 × 84 squares are shown in blue and orange. The 84 × 24 rectangle is shown with the black perimeter. The three 24 × 24 squares are shown in red, green and purple. The two 12 × 12 squares are shown in yellow and white.

24 **(C) 10**

Any time there are 3 or more ties, we can increase the number of losses by replacing 3 tied games by 1 winning game and 2 losing games. The number of games is still the same since 1 + 2 = 3 and the number of points is also the same since 3 + 0 + 0 = 1 + 1 + 1. For the greatest number of losses, the number of ties must be either 0, 1, or 2.

If there were no ties, then 80 points would be equal to 3 points × the number of wins. This cannot happen since 80 is not a multiple of 3.

If there were only one tie, then 80 − 1 = 79 points would be equal to 3 points × the number of wins. It cannot happen since 79 is not a multiple of 3.

The only option left is 2 ties, so the number of wins is 26 since 3 × 26 + 2 = 80. In this case, the number of lost games is 38 − (26 + 2) = 10, so 10 is the greatest possible number of games that the team lost.

2014

3 Point Solutions

1 **(D)** 

The large figure with the star has 9 spikes. In the picture below the number of spikes in each figure was counted. The only figure with 9 spikes is (D), so it is the only figure that could be the central part of the larger figure.

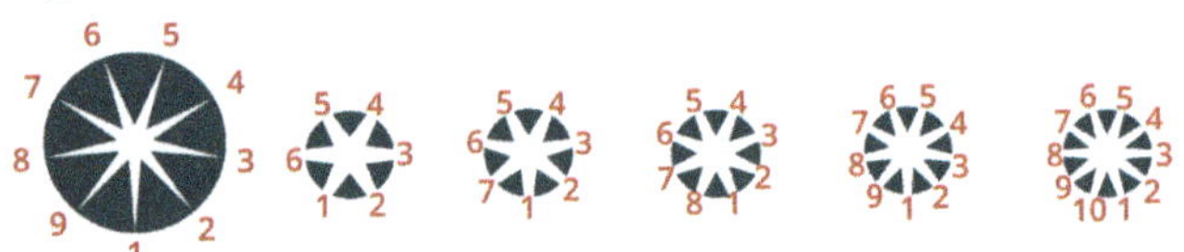

2 **(D) between the 1 and the 4**
To make the smallest number possible, the smaller digits need to come before larger digits. The digits 2, 0, and 1 are smaller than 3, so 3 should be placed after those digits. 4 is larger than 3, so 3 should be placed before 4. So, Jackie should place the digit 3 between the 1 and the 4 to make the number 20,134, which is the smallest number she can make.

3 **(A) 1, 4**
House 5 has a brown triangle roof different from any other house, House 2 is the only house made with a smaller yellow square, and House 3 is the only house with a small blue rectangle. Each of these three houses is made with different pieces than any other house. Only houses 1 and 4 are made using exactly the same triangular and rectangular pieces.

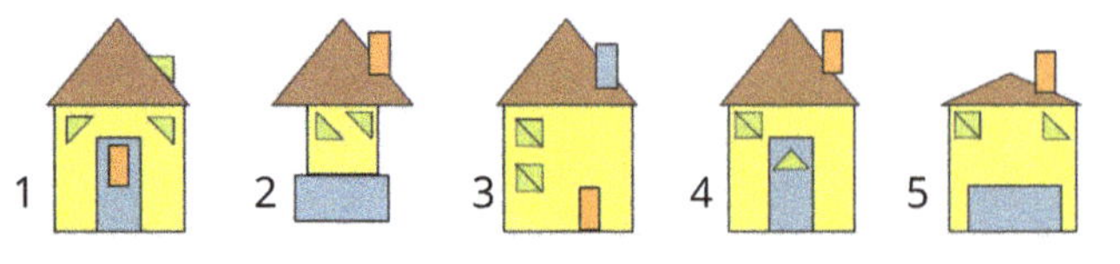

4 **(D) 200**
Yesterday Koko was awake for 24 − 20 = 4 hours. Since he eats 50 gram of leaves per hour, during the 4 hours he was awake he ate 50 × 4 = 200 grams of leaves.

5 **(A)**

The results of all the subtractions performed by Maria are marked in green. The red line connects subtraction results from 0 to 5, in increasing order. The figure that Maria draws is seen in (A).

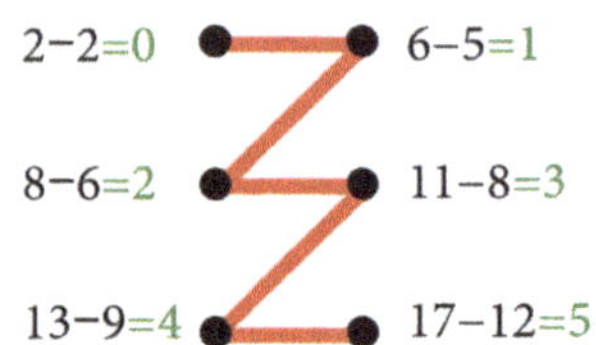

6 **(E) Lucy**
In the picture below, the names of the children are lined up according to the number of sandcastles they built. Of all 5 children, Lucy built the most sandcastles.

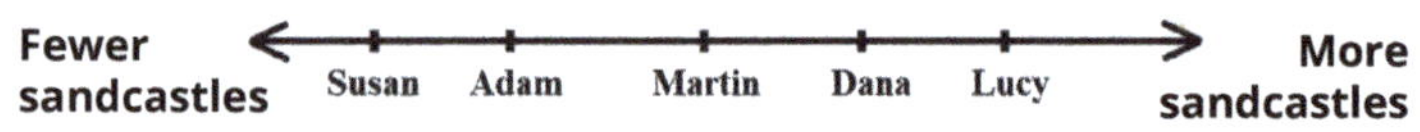

7 **(E) 8**
The picture below shows all the steps needed to find the number Monica wrote in the gray cell. The number written in the gray cell is 8.

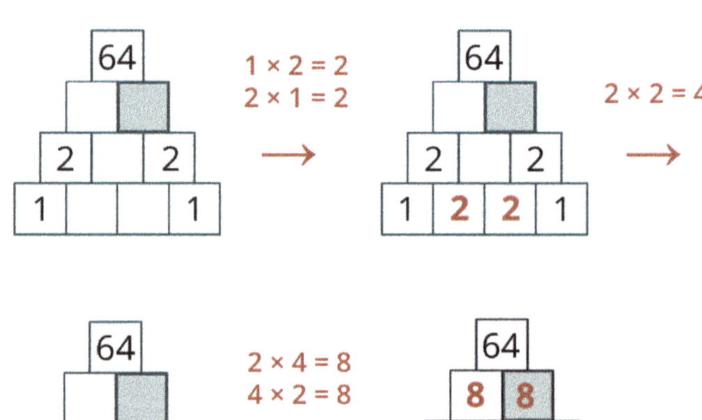

8 **(C)** 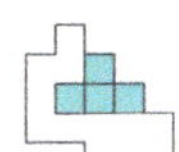

Each of the four pieces is made of 4 small squares. The longest piece is 4 small squares long. The white shape has two columns and one row (shown below in red) of exactly that length. In the first case, there is not enough area to the left of the red piece to place there any piece made of four small squares. In the second case, the area to right has 5 small squares, more than needed for one piece but not enough for two pieces. The third case looks more promising. The T-shaped piece fits exactly in the area above the red row and the area below has 4 + 4 small squares. However, the two pieces

cannot fit together in the bottom part of the white shape.

There are two other options to place the longest of the four pieces, shown in red, in the second row from the bottom. The first option shows only 3 small squares below the red piece, so no other piece can fit there. The second option shows below the long red piece.

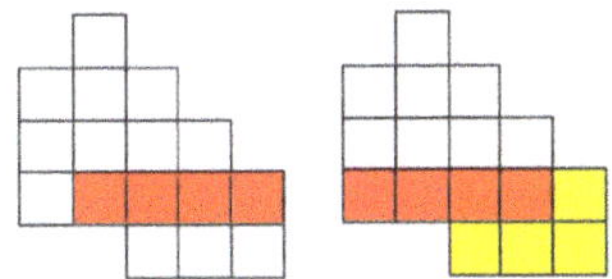

There is only one way to place the other two pieces as shown. The T-shaped piece is placed in the position shown in (C).

4 Point Solutions

9 **(E)**

From the other side of the window, the figure will look like its mirror image. The picture shows the original painting and its mirror image, which is the flower seen in (E).

Mirror Image

---

10 **(E) 48**

In the end there were 6 pieces of candy left in the bowl, so Clara took 6 pieces out, which is half of the 12 pieces that were there before. Tom took 12 pieces of candy out of the bowl, which is half of the 24 pieces that were there beforehand. Sally took 24 pieces out of the bowl, half of the 48 pieces that were there at the beginning. This process is also shown in the diagram below.

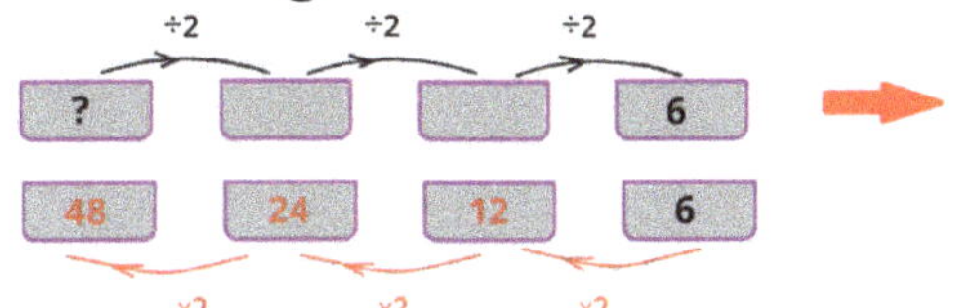

---

11 **(B)** 

The original picture with one tile missing has 3 tiles exactly half dark gray and half light gray. There are 2 dark gray tiles matched by 2 light gray tiles and 1 light gray tile left over. That left over light gray tile must be matched by one full dark gray tile which is the one that must be added to the picture.

12 (D) 90

When missing the target both times, Paula gets 0 points. When hitting the target only once, she can get 30, 50, or 70 points. If she hits the same area twice, Paula can get 60, 100, or 140 points. If she hits two different areas, Paula can get 80, 100, or 120 points. From the list of the sums given in the answers, only 90 points cannot happen.

---

13 (B) 6

We can see 4 yellow tokens, 4 green tokens, and 5 red tokens (one is in the middle). Altogether 4 + 4 + 5 = 13 tokens are visible. Mary also has 5 tokens that were not used, so at the beginning she had 13 + 5 = 18 tokens. Since she had equal numbers of tokens in each of the 3 colors, she had 18 ÷ 3 = 6 tokens of each color, so Mary had 6 yellow tokens at the beginning.

---

14 (B) 7

Peter Rabbit ate 30 carrots in a week by eating either 9, 4, or 0 carrots per day. If he ate only 4 carrots per day, he could not eat 30 carrots during the week since 30 is not a multiple of 4. If he ate only 9 carrots per day, he could not eat 30 carrots during the week since 30 is not a multiple of 9. Therefore, there was one day when he ate 4 carrots and another day when he ate 9 carrots. During the other five days he ate the other 30 − 4− 9 = 17 carrots. 17 is not a multiple of 4 and 17 is not a multiple of 9, so there were two other days when Peter Rabbit ate 4 + 9 = 13 carrots. He had to eat 17 − 13 = 4 more carrots on another day. Thus, there were two days when Peter Rabbit ate 9 carrots per day and no cabbages, and three days when he ate 4 carrots and one cabbage each day. There were also two days when he ate 2 cabbages each day and no carrots. So, during the week Peter Rabbit ate 2 × 0 + 3 × 1 + 2 × 2 = 7 cabbages.

15 (C) 

From directly above, only the sides marked with red and green dots will be visible. So, when looking directly from above the solid will look like shape (C).

Side View Top View

---

16 (B) 181

We can color squares in the 4 non-overlapping rows. Each of these rows has exactly 8 squares, so there are 4 × 8 = 32 colored squares. Each of them has 5 dots, so there are 32 × 5 = 160 dots inside the colored squares. There are 3 rows of dots between the colored rows, each with 7 dots, so there are 3 × 7 = 21 such dots. The total number of dots in the picture is 160 + 21 = 181.

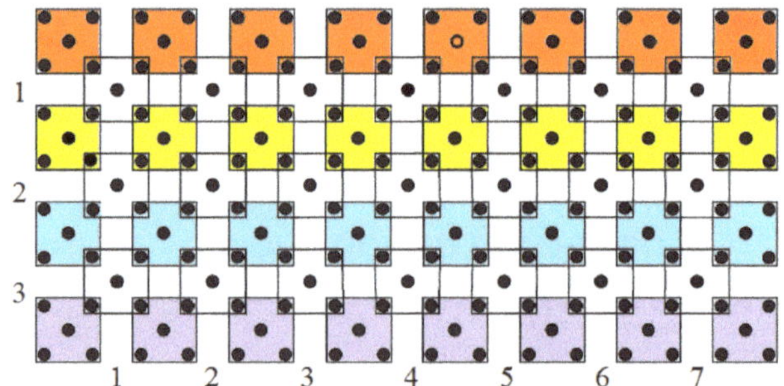

5 Point Solutions

17 (B) 30

A kangyear has 20 kangmonths, so there are 5 kangmonths in one quarter of a kangyear since 4 × 5 = 20. Each kangmonth has 6 kangweeks, so there are 5 × 6 = 30 kangweeks in one quarter of a kangyear.

18 **(C) 4 is the only possible number.**
A group of 7 girls standing in a circle violates the rule that no three girls are standing next to each other. Same is true for a group of 6 girls and one boy. To avoid another violation of the above rule, a group of 5 girls must be split by two boys into two smaller groups. One of these smaller groups must have at least 3 girls, again violating the rule that no three girls are standing next to each other.

2 or more boys standing as part of the circle divide the circle into as many parts as the number of boys. No two boys are standing next to each other, so there is at least one girl in each part of the divided circle. Thus, the whole group of children must have at least as many girls as boys. In the case of 7 children, the number of boys is at most 3, so the number of girls is at least 7−3 = 4. We already know that more than 4 girls would violate the rule that no three girls are standing next to each other.

In conclusion, the only possible option is that in the group of seven children exactly 4 are girls, so 4 is the only possible number. The picture shows 4 girls and 3 boys standing in the circle according to both rules.

19 **(B) 3**
Only two cards are in their right places, so 6 cards need to be moved. In one move we can switch only two cards, so at least 3 moves are needed to switch 6 cards. Indeed, it can be done in 3 moves as shown below.

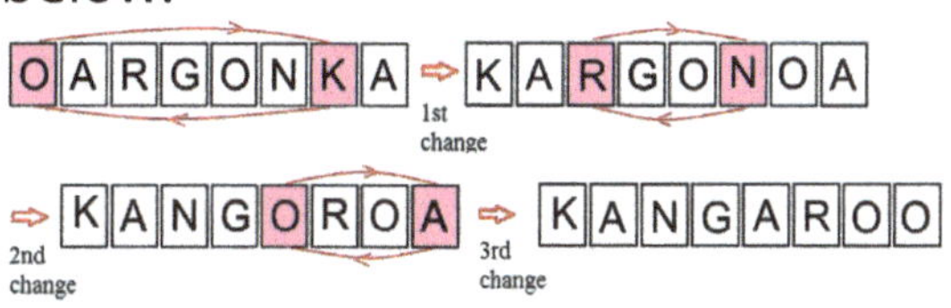

20 **(C) 26**
All 6 steps are shown in the picture below. The equations show the total number of diamonds in the figures above them. There are 3 diamonds used in step 1. In step 2, the bottom row with 3 diamonds was added. For step 3, the bottom row with 4 diamonds was added. The pattern continues. Step 6 will consist of 28 diamonds with 2 white diamonds and 26 black diamonds.

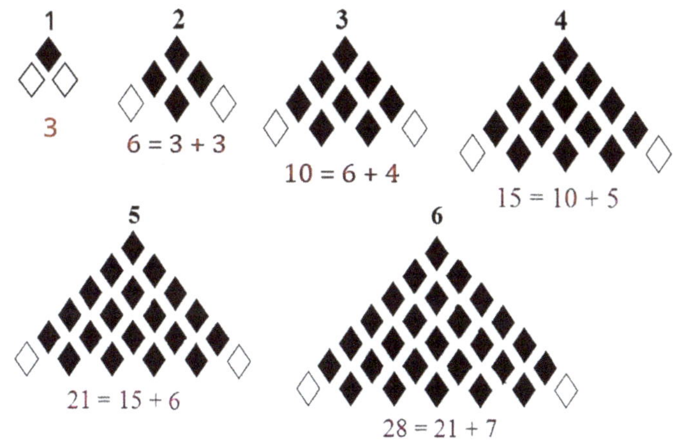

21 **(A) the bear and the horse**
At first, Hamish bought some toys for 150 − 20 = 130 Kangcoins. For three toys he would pay at least 40 + 48 + 52 = 140 Kangcoins, so he bought exactly two toys. Only 73 + 57 = 130 Kangcoins, so at first Hamish bought the bear and the duck. After the exchange he got back 5 Kangcoins, so he exchanged one of his toys for another toy that was 5 Kangcoins cheaper. Only the horse is 5 Kangcoins cheaper than the duck, so Hamish left the store with the bear and the horse.

22 **(D) 5**

When we add two two-digit numbers with all different digits that are 6 or less, the sum is at most 117, since 64 + 53 = 63 + 54 = 117. So, the hundreds digit of the sum is 1 and its tens digit has to be 0, since 1 cannot be used twice. The two addends (numbers being added) can use only the remaining digits 2, 3, 4, 5, and 6. Using only 2, 3, 4, and 5, would make the sum at most 95 since 53 + 42 = 95 or 52 + 43 = 95. This sum would never be a 3-digit number. Thus, 6 has to be a digit of one of the addends. 6 cannot be the ones digit of either addend since 6 + 2 = 8, 6 + 3 = 9, 6 + 4 = 10, and 6 + 5 = 11. The ones digits of these results are 8, 9, 0, and 1. However, 0 and 1 are already used, and 8 and 9 are not on the list of digits. For the same reason 5 cannot be the ones digit of either addend. 5 and 6 cannot be the tens digits at the same time since the sum would be greater than 110 but we already know that the tens digit of the sum is 0. Thus 5 has to be placed in the shaded square. Ignoring the order of the addends, the possible solutions are displayed below.

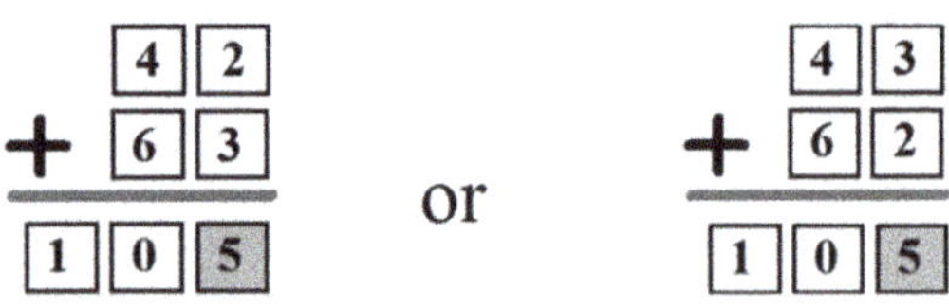

23 **(D) 21**

There are four 2 × 2 non-overlapping squares shown in different colors in the figure on the left. Each of them must have at least one unshaded unit square, so the figure must have at least 4 unshaded small squares. The figure has 25 small squares, so we cannot shade more than 25 − 4 = 21 of them. As shown, it is possible to shade exactly 21 small squares.

24 **(D) 8**

The shaded cell has 4 neighboring cells. In these cells we can put only numbers greater than 4. The sum of numbers in these 4 neighboring cells is never 13 since it is greater than 4 × 4 = 16. Therefore, neither 5 nor 6 can be written in the shaded cell. Only 7, 8, or 9 can be written in that cell.

Each of the four middle edge cells has three neighboring cells. In each case, the sum of numbers in the three neighboring cells depends on the number in the shaded cell. When 7 is written in the shaded cell, the sums of the middle and corner cells that would be the neighbors for 5 or 6 are:

1 + 2 + 7 = 10, 2 + 3 + 7 = 12,
3 + 4 + 7 = 14, and 4 + 1 + 7 = 12.

None of the results is 13, so 7 cannot be written in the shaded cell.

If we replace 7 by 9, then the corresponding sums are:

1 + 2 + 9 = 12, 2 + 3 + 9 = 14,
3 + 4 + 9 = 16, and 4 + 1 + 9 = 14.

Again, none of the results is 13, so 9 cannot be written in the shaded cell. 8 is the only option left and the corresponding sums are:

1 + 2 + 8 = 11, 2 + 3 + 8 = 13,
3 + 4 + 8 = 15, and 4 + 1 + 8 = 13.

Two of the sums are equal to 13, so 5 and 6 (in any order) can be written in the cells between 2 and 3 and between 4 and 1.

In either case, Nick wrote 8 in the shaded cell.

1		2
5	8	6
4		3

or

1		2
6	8	5
4		3

2016

3 Point Solutions

1 **(E) Ernst**
The totals are:
Amy: 6 + 1 = 7
Bert: 3 + 3 = 6
Carl: 2 + 3 = 5
Doris: 4 + 4 = 8
Ernst: 5 + 4 = 9

The largest total is 9. This is what Ernst rolled.

---

2 **(E) 5**
8 weeks = 7 weeks + 2 days + 5 days, so Little Kanga will be 8 weeks old in 5 days.

---

3 **(A) 24**
17 + 3 = 20, 20 − 16 = 4, and
20 + 4 = 24.

---

4 **(A)**

An experiment shows a mirror image of
F as ꟻ

so objects are reversed left-to-right but not top-to-bottom. It is like looking at a flat picture on the front of a transparent paper from the **back** of the paper.

---

5 **(D)**

71 and 72 are seats between 61 and 80. It is indicated by

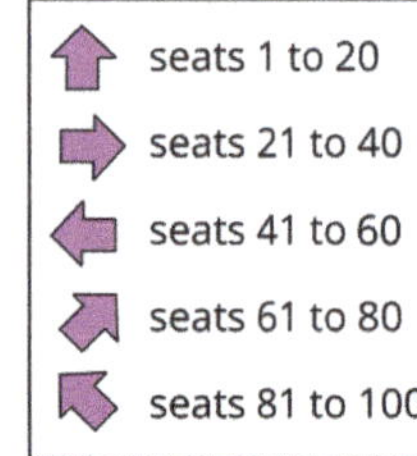

6 **(B) 3**
There are 6 people involved, Anna + 5 friends. 6 × half of an apple = 3 apples.

---

7 **(A) a triangle**
Extend the shaded region to a full rectangle to see a triangle behind the curtain.

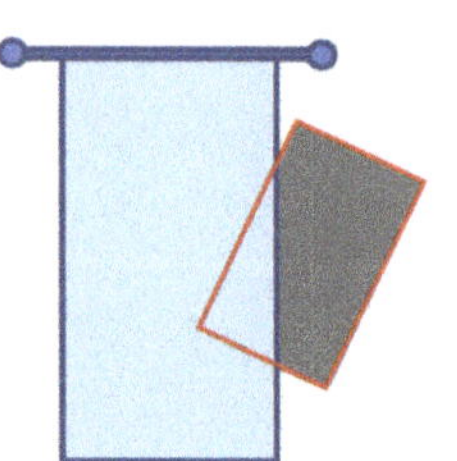

---

8 **(C) There are twice as many circles as triangles.**
There are 4 circles, 2 squares, and 2 triangles. 4 = 2 × 2, so there are twice as many circles as triangles which makes the statement (C) true. The other statements are false.

4 Point Solutions

9 **(B) 2025**
If the sum of digits of any year is 9, then the year must be a multiple of 9. 2016 is a multiple of 9, so we look at the next multiple of 9, 2016 + 9 = 2025, and check that 2 + 0 + 2 + 5 = 9. (Be aware that we were just lucky that the very next multiple worked. If the starting year was, for example, 2070, then the next multiple of 9 would be 2070 + 9 = 2079, with the sum of the digits equal to 18, not 9. The first year after 2070 with the sum of its digits equal to 9 will be 2106.)

10 **(B) 4**

The mouse can't go back through any gate, so it must pick (in the first move) one of the purple gates and then one of the green gates. 2 × 2 = 4, so there are 4 different path choices (yellow, black, blue, and red).

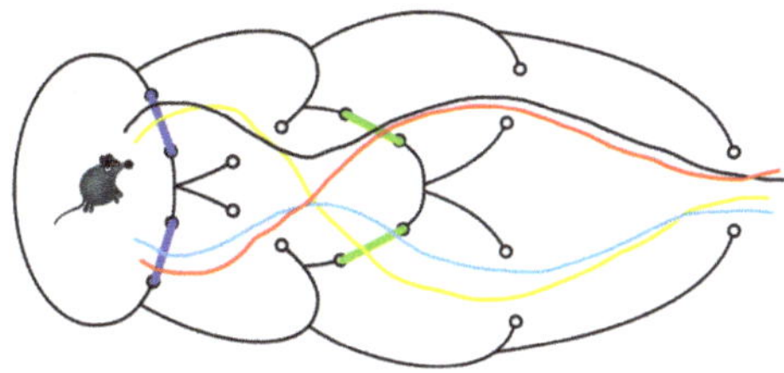

11 **(C) 11 and 4**

On each of two cards Zoe has a sum of two numbers. By adding them, we are adding all four numbers, so the total of the two sums is 32. These two sums are equal, so the sum on each card is 32 ÷ 2 = 16. The numbers we cannot see are 16 − 5 = 11 and 16 − 12 = 4.

12 **(B)**

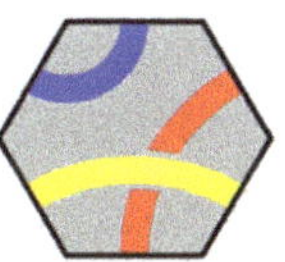

Blue is the only path which connects adjacent sides of the middle tile. This means that only tiles (B) or (C) can possibly work. Notice that in the picture tile (B) fits directly in the middle, whereas even after rotating of tile (C) so that the blue path in the upper left corner, the red and yellow paths are switched.

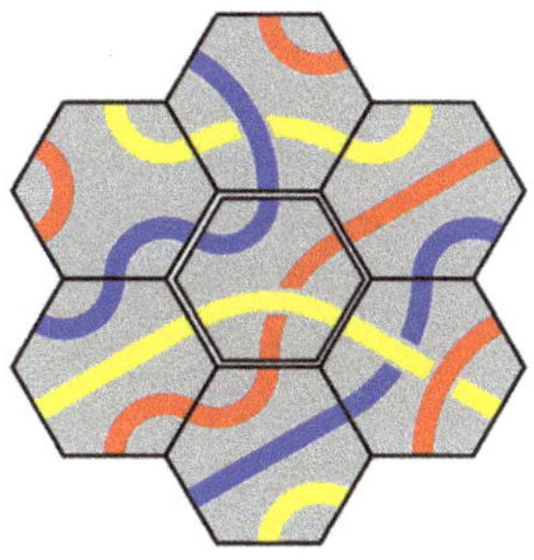

13 **(D) 3**

The circle doesn't cover any vertex of the square or the triangle. If four vertices of the square are visible, then the square is above the triangle. Otherwise, the triangle is above the square. Thus, the 1st pile, the 4th pile, and the 5th pile are all piles with the triangle above the square.

14 **(A) 1, 3, and 5**

The bottom of 1 has the same length as the bottom of 3, so they can be opposite sides of the square. Fit 5 between the two pieces to form a square.

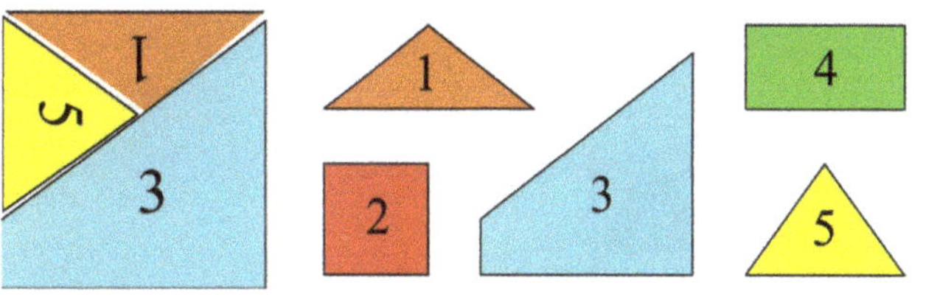

15 **(C) 4**

There is only one way to fill in the table. Here are the first steps:

1		
	2	■
		■

⇒

1	3	
	2	■
		■

⇒

1	3	2
	2	■
		■

We now see that 1 and 3 need to be in the two shaded cells, so the sum is 4. The solution for the whole table is shown below.

1	3	2
3	2	1
2	1	3

16 **(D) 5**
Only the first three cells and the last three cells can be empty, so 11 − 3 − 3 = 5 cells must each contain a coin.

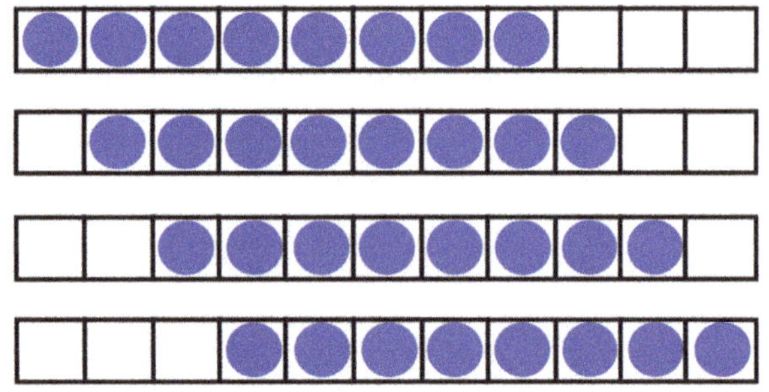

5 Point Solutions

17 **(D)**

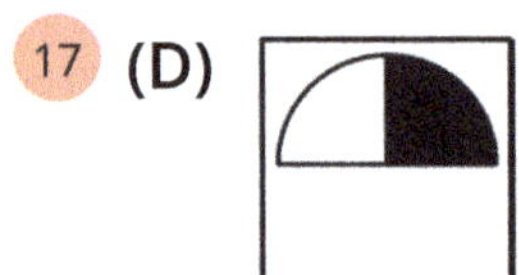

Each turn is like a mirror reflection. Turn over the card on the lower right on its left edge to see the picture of the original card in the lower left. Next, turn over the card in the lower left on its top edge to see the picture on the top.

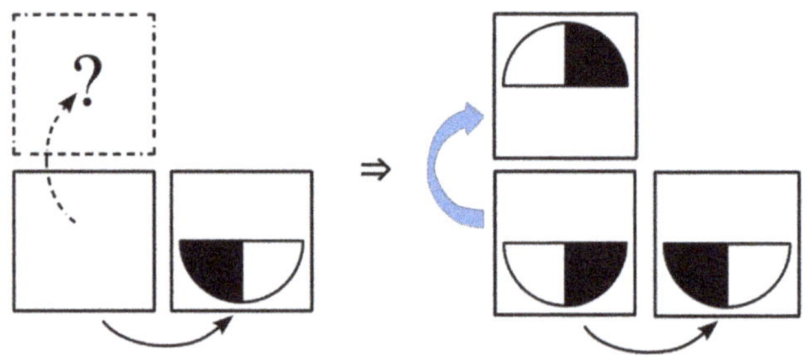

18 **(B) 27**
If we subtract 3 from the unknown sum of the four brothers' ages, then we have a multiple of 4 since Paul's age − 3 is the age of the other three brothers. From the list of 25 − 3, 27 − 3, 29 − 3, 30 − 3, and 60− 3, only 27 − 3 = 24 is a multiple of 4. Thus, 27 can be the sum of the ages of the four brothers: 6 + 6 + 6 + 9 = 27.

19 **(D) 50**
Each tree has twice as many pears as apples. Thus, any magic garden with 25 apples must have 2 × 25 = 50 pears. There are two types of magic gardens with 25 apples. We may have 7 trees with 3 apples each and 1 tree with 4 apples, or we may have 3 trees with 3 apples each and 4 trees with 4 apples each. In short, 7 × 3 + 1 × 4 = 25 and 3 × 3 + 4 × 4 = 25.

One garden is shown:

20 **(C) 6**
Each dog has 3 more legs than noses, so 6 dogs have 6 × 3 = 18 more legs than noses.

21 **(B) between bowl Q and bowl R**
Q weighs less than R. Remove the square from both to see that one triangle weighs less than one circle.

Q weighs less than Z since one triangle weighs less than one circle. For the same reason, Z weighs less than R.

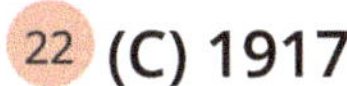

(C) 1917

The sum of the six numbers, without 201, is 2016 − 201 = 1815.
The sum of seven numbers, with 102, is 1815 + 102 = 1917.

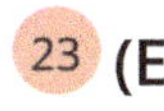

(E) 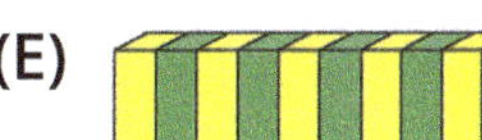 **10**

27 = 9 + 18, so the long bar is broken into bars of 9 and 18 bricks.

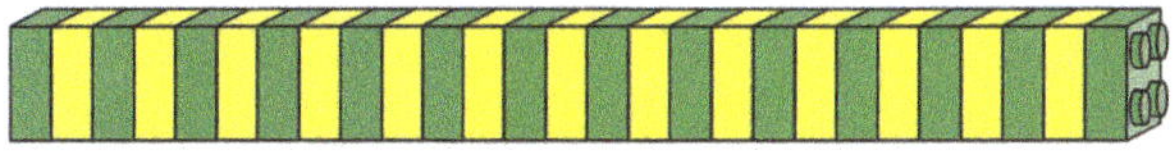

We cannot get the bar of 10 bricks from the bar of 9 bricks.

 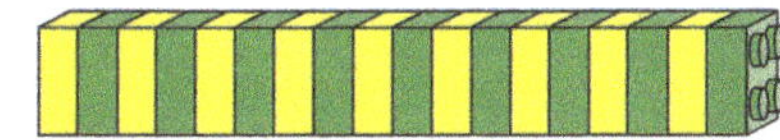

18 = 6 + 12, so the only chance to get a bar of 10 bricks is from the bar of 12 bricks, but 12 = 4 + 8, so we cannot get a bar of 10 bricks. We already know how to create bars of 8, 6, and 4 bricks. 6 = 2 + 4, so we also can get a bar of 2 bricks.

24 **(B) Bertha**

Originally, Angel chirps 4 times, Bertha once, Charlie twice, David 3 times, and Eglio 4 times.

If Angel turns, he would not chirp at all. If it was Bertha, she would chirp 3 times.

If Charlie turns, he would still chirp twice. If David turns, he would chirp only once.

If it was Eglio, he would not chirp at all.

So, only Bertha chirps more times when she looks in the opposite direction.

See the picture:

2018

3 Point Solutions

1 **(B) 6**
Anytime a digit is repeated, the same stamp is used. So, only count the number of different digits in the date.

2 **(E) 6**
The picture shows the path of the arrows. The balloons which each arrow hits are marked with an X.

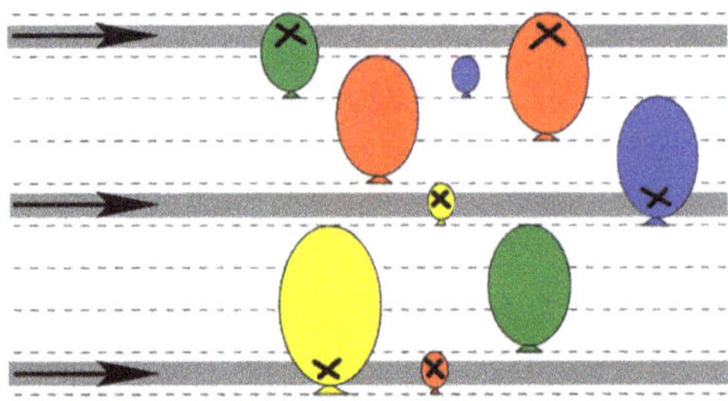

3 **(C) 18**

Susan
6 years old

Sister
(6 – 1) years old

Brother
(6 + 1) years old

$6 + 6 + 1 + 6 - 1 = 6 + 6 + 6 = 18$

4 **(E) 5**
If screw 5 was as long as any of the other screws, the bottom would be visible under the block like it is for screw 1. Therefore, screw 5 is the short screw.

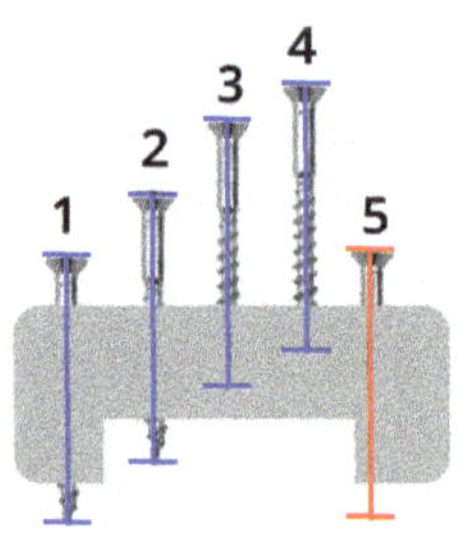

5 **(D)** 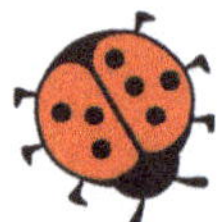

Sophie has 3 dots on her left wing and 4 dots on her right wing. The ladybug in answer choice (D) has 4 dots on her left wing and 3 dots on her right wing. Therefore, (D) shows a different ladybug. The other four answer choices match the dots Sophie has.

6 **(D)** 

The folded sheet of paper has a large opening that looks like a half-circle. If the paper is opened and this is the middle, there will be a hole resembling a circle in the middle of the sheet.

7 (D) 21

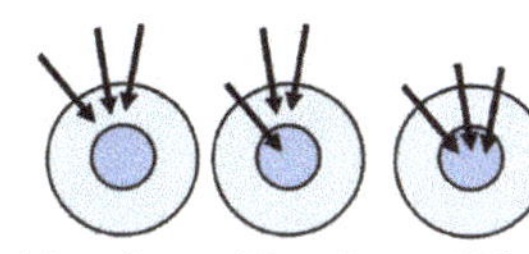

12 points 15 points ???

12 points = 4 + 4 + 4
15 points = 4 + 4 + 7
7 + 7 + 7 = 21 points
On the first turn, each of the arrows hit an area with the same color, so Diana received the same number of points for each arrow. Since there were three arrows, Diana scored 4 points with each of them (4 + 4 + 4 = 12). On the second turn, one of the arrows landed on an area worth more points. Diana scored 4 + 4 = 8 points with the two arrows on the lighter blue area, so the darker blue area is worth 15 – 8 = 7 points for each arrow that lands there. So, on the third turn, Diana scored 7 + 7 + 7 = 21 points.

---

8 (A) 5
Going around the table, think of the right and left side as a person facing it from that direction would see it.

4 Point Solutions

9 (D) 4
The original tile has one circle and one star, so any design must have an equal number of circles and stars. The design shown second from the left has six circles and eight stars, so Roberto definitely cannot make it. Below are suggestions how the other designs can be made.

10  

The two figures missing in the fourth column are the rhino and the ghost. If Albert places the rhino in the spot with the question mark, then he would have to place the ghost in the only other empty spot in the fourth column, making two ghosts in the second row, which is not allowed. Therefore, the ghost must go in the spot with the question mark.

---

11 (B) 6
Tom needs to use the squares for the top part of the boat because of the right angles. Three squares will fill the top part of the boat. Three trapezoids will make up the bottom of the boat if two are flipped.

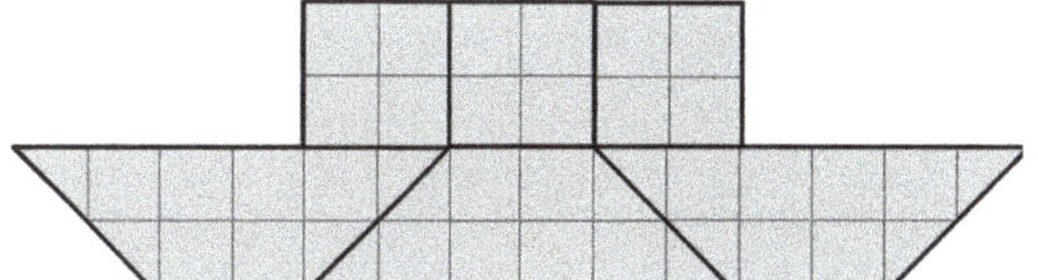

---

12 (E)

Only choices (C) and (E) have the right number of dots. However, in (C) the black dots are too close to the one that is outlined.

13 **(E) Friday**
Twelve carrots take 6 days to eat. Work backwards with the days of the week.
11th and 12th carrots: Wednesday
9th and 10th carrots: Tuesday
7th and 8th carrots: Monday
5th and 6th carrots: Sunday
3rd and 4th carrots: Saturday
1st and 2nd carrots: Friday

---

14 **(C) 8**
Each cube has 6 faces. Where it touches another cube, a face is covered, and will not be painted. So, any cube that touches exactly two other cubes will have exactly four faces exposed and painted. This is true for all cubes in this figure except for two, one at each end of the structure. So, 10 – 2 = 8 cubes will be painted on exactly 4 faces.

---

15 **(C) 4**
The number of butterflies on the flowers is twice the number of dragonflies on the flowers, so the total number of insects on the flowers is 3 times the number of dragonflies on the flowers. That total is between 5 and 8 since the total is more than half or 8, which is 4, and at most 8 since no more than 1 insect is on each flower. The only multiple of 3 between 5 and 8 is 6, so there are 6 insects and one-third of them (which is 2) are dragonflies. Hence, 6 – 2 = 4 insects are butterflies.

---

16 **(D) 33 km**
Because the distance between Easter and Flower going through Volcano is 26 km, we can disregard the information that the distance between Easter and Volcano is 17 km. Now we can subtract the two distances of 26 km each and the 15 km between Flower and Desert from the 100 km round trip to find the distance between Easter and Lake. 100 – (26 + 15 + 26) = 100 – 67 = 33.

5 Point Solutions

17 **(D) D**
Two paths ending at the door D are shown. There is no path ending at any other door. Otherwise such a path could be reversed and the reversed path would show decreasing numbers. Starting at A we could move from 12 to 9 to 5 (in two ways) and not farther. Starting at B we could move from 14 to 12 to 9 to 5 (in two ways), from 14 to 5, or from 14 to 6 and not farther. Starting at C there is no way to move anywhere from 6. Finally, starting at E there is no way to move from 9.

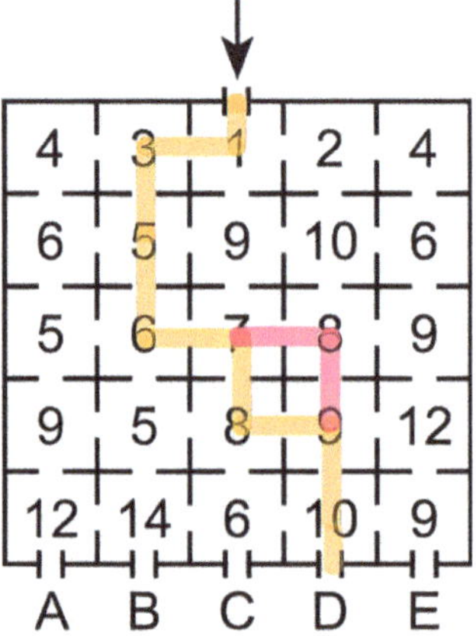

---

18 **(C) C**
According to the second scale, C weighs **at least** 10 g + 20 g = 30 g, so C and D together weigh at least 30 g + 10 g = 40 g. According to the first scale, C and D together weigh less than half of all four balls together. The four balls weigh 100 g since 10 + 20 + 30 + 40 = 100, so C and D together weigh less than 50 g. Hence, C and D together weigh **at most** 40 g.

Therefore, C and D together weigh **exactly** 40 g, C alone weighs 30 g, and D alone weighs 10 g.

Notice that B weighs 30 g – 10 g = 20 g and A weighs 40 g.

19 (B) 8 cm

Tightening the band by one hole makes it 2 cm shorter. It does not matter that there are two layers in the overlap, because you can think of the lower layer as not moving. Moving from one hole to five holes decreases the length by 4 × 2 cm = 8 cm.

---

20 (A)

The result of adding the same number four times can be 4, 8, 12, 16, ...

+ =

and

+ =

so

+ + + =

Among the numbers 1, 2, 3, 4, and 5, only 4 is a possible result, so

= **4**, = **1**, and

= **2** **1+4=5**, so **=5**

Hence, represents the number 3.

---

21 (B)

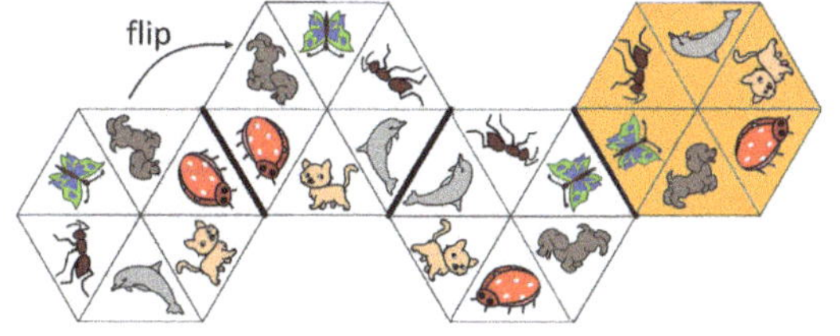

---

22 (D) 198

The smallest squares have an area of 1, and therefore a side length of 1, because 1 × 1 = 1. Knowing this, we can find the side lengths of all the squares that make up the large rectangle. The side lengths of the rectangle are 18 and 11, so the area is 198. We can also add the areas of all the squares to find the area of the rectangle.

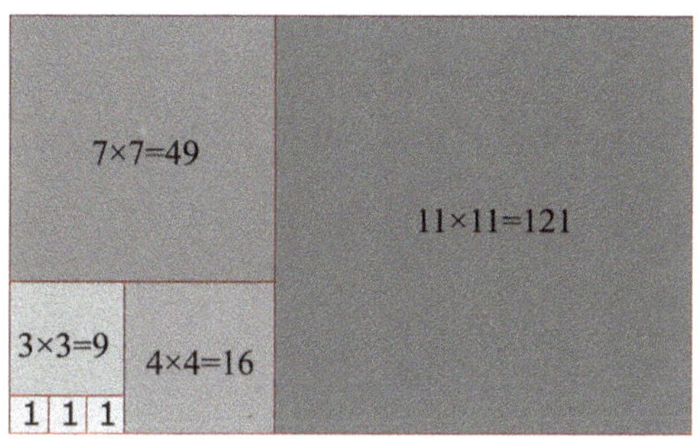

---

23 (E) only the numbers 1 and 7

The cell with the question mark has 5 neighbors, and there is only one cell that is not a neighbor with it. A number in the sequence that cannot be placed next to each other will be one more or one less than a given number. Because there are 7 numbers, the only numbers that can be written in the cell with the question mark are the numbers either at the start or at the end of the sequence, because they have only one other number in the sequence that cannot be placed next to them. These are the numbers 1 and 7. Placing any other number in the cell with the question mark will violate the rules because then consecutive numbers will necessarily be neighbors with it.

---

24 (B) 9

Each time 3 heads are cut off, 1 new head grows. 13 heads = 4 × 3 heads + 1 head, so the dragon grew 4 new heads. Together with the original heads, there were 13 heads cut off.

Therefore, 13 − 4 = 9 is the number of heads the dragon had at the beginning.

2020

3 Point Solutions

1 **(E)** 

The smallest picture is in answer (B), which represents a photo of the mushroom taken on Monday. The next in height is in answer (E), so (E) represents a photo of the mushroom taken on Tuesday.

---

2 **(E)** 

The picture that completes the pattern completes two 4-ray pointy shapes across from each other, one 4-ray round shape and one 4-ray indented shape. This is shown only in piece (E).

---

3 **(A)** 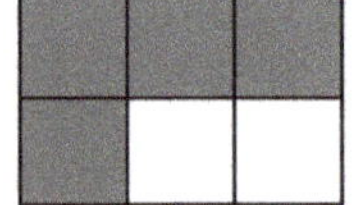

The calculations are as shown. After shading, the shape looks like answer (A).

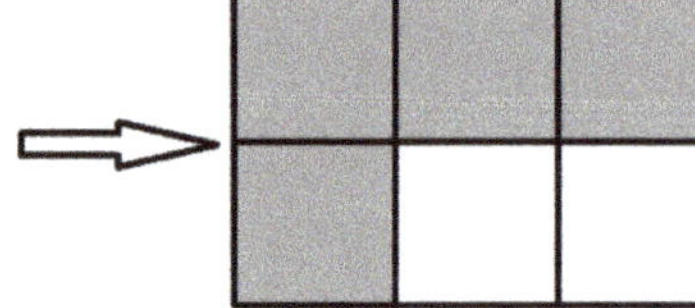

$16+4$ $=20$	$19+1$ $=20$	$28-8$ $=20$
2×10 $=20$	$16-4$ $=12$	7×3 $=21$

4 **(A)** 

It is easier to see the size of the region that is not shaded by counting the white squares (or 2 halves that make one whole square). Figure (A) has an area of 3 squares that is not shaded; figure (B) has an area of 3 and a half squares that is not shaded; figure (C) has an area of 3 and a half squares not shaded; figure (D) has an area of 3 and a half squares not shaded; figure also (E) has an area of 3 and a half that are not shaded. Thus, figure (A) has the largest part shaded as it has the smallest part NOT shaded.

---

5 **(E)** 

There are 6 shapes given: 2 identical circles, 3 different-sized triangles, and 1 narrow rectangle. Notice:

(A) shows two top triangles of the same size.
(B) shows a square that is not given.
(C) shows one circle of a different size.
(D) shows two triangles of the same size.
(E) shows exactly the same six shapes as those given.

---

6 **(D) 19**
The red arrows below show Elli's jumping pattern. The largest number that Elli can reach by jumping to a number that is 3 more than the previous one is 19.

1	5	8	11
4	7	10	14
24	23	13	18
21	19	16	20

7 **(E)**

This solution is obtained by eliminating those stickers that are not opposite the duck. From the first cube we observe that the mouse and the ladybug are not on the face opposite the duck. From the second cube we observe that the elephant and a dog are not on the face opposite to the duck. This leaves us with only one remaining possibility for the face opposite the duck: the sticker of the fly.

---

8 **(C) 5**

The pieces consist of 1, 2, 3, 4, 5, 6, and 7 squares and the grid consists of 17 squares. Certainly we cannot use all the pieces as 1 + 2 + 3 + 4 + 5 + 6 + 7 = 28, which is more than 17. Also, we cannot remove just 1 piece as the total length would then be at least 1 + 2 + 3 + 4 + 5 + 6 = 21, which is still larger than 17. This leaves us with the need to remove two or more pieces, which means we can use at most 7 – 2 = 5 pieces. It can be observed that by removing the pieces of length 5 and 6 squares, the rest covers the grid completely, as 1 + 2 + 3 + 4 + 7 = 17. Figure below demonstrates how the pieces fill out the space. The greatest number of pieces we can use is 5.

4 Point Solutions

9 **(C) 3**

Since the outer region is red, the following two going towards the center are blue and yellow. The next would have to be red again, followed by blue and yellow. The center piece would be red again. This gives us three pieces colored red, and no areas of red touch. Notice that blue and yellow are interchangeable colors.

---

10 **(C)**

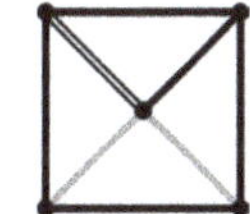

Notice that two solid light gray edges of the pyramid are next to each other. This eliminates answers (A), (B), and (E). If Loes stands in front of the base edge and looks at the pyramid from above, she would see the following pattern:

Rotating 180 degrees would show the pyramid as in answer (C):

---

11 **(D) 4**

The dog's leash is 11 units long. Following the hut as close as possible to the wall, the dog will be able to reach 4 bones as shown in the picture.

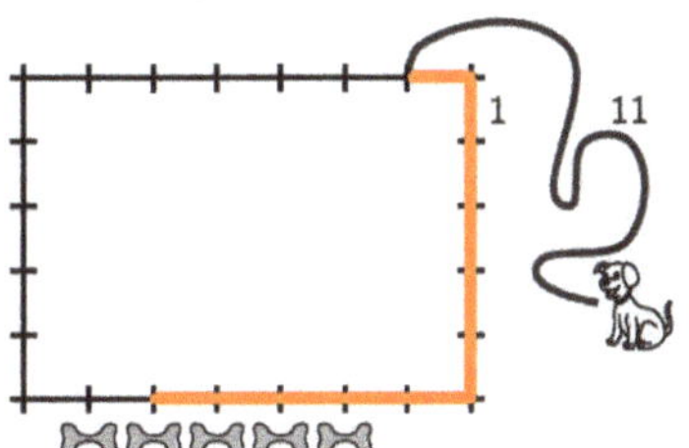

12 (E) 42

Notice that 10-meter-long fence would have to have 10 upper horizontal poles, 10 middle horizontal poles, 11 vertical poles in the top row and 11 vertical poles in the bottom row. The total number of poles would be 10 + 10 + 11 + 11 = 42.

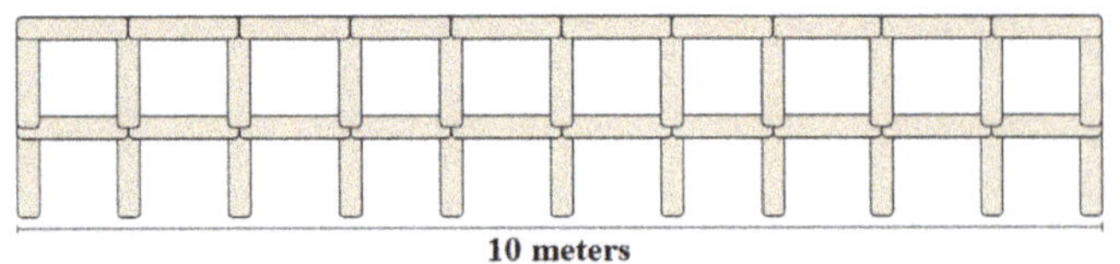

---

13 (D) 70

From the information that every time the kangaroo goes up 7 steps, the rabbit goes down 3 steps, we conclude they move together a total of 7 + 3 = 10 steps. As there are 100 steps, they need 100 ÷ 10 = 10 moves until they meet. At that point the kangaroo has gone up 7 × 10 = 70 steps.

---

14 (A) 9

The sum of the numbers after the secret number was subtracted is 24 + 13 + 7 = 44. The original sum of the numbers was 50. The difference between 44 and 50 is 6. This makes the sum of three of the same number be 6. Therefore, the secret number is 2, as 2 + 2 + 2 = 6. The original numbers are: 24 + 2, 13 + 2, and 7 + 2, or 26, 15, and 9. From the numbers listed in the answers, the correct choice is answer (A) 9.

15 (D) 4

Notice that the token is divided into five equal triangles. The numbers 1 to 5 are written on the token counterclockwise. Following the rules that the parts of the token touching have the same number and the numbers on each token's triangle are written counterclockwise, we get the pattern shown in the picture. Number 4 would go on the triangle marked with the X, here highlighted in yellow.

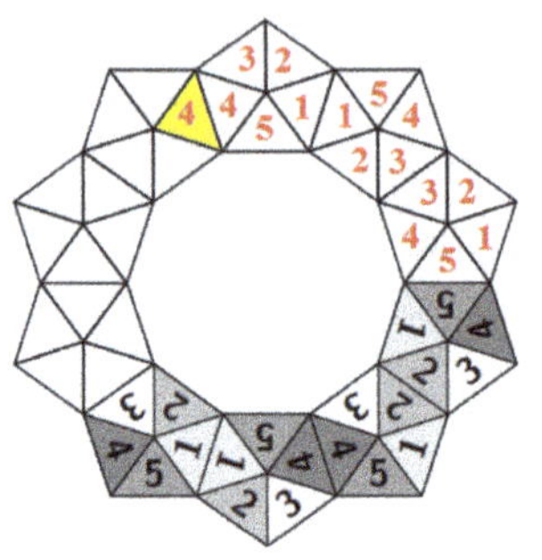

---

16 (B) 3 short and 3 long

(A) Five short sticks cannot make two sides both equal to 3 sticks.

(B) Since the short stick is three times shorter than the long stick, three short sticks are equal in length to one long stick, making it possible to construct a square.

(C) Six short sticks cannot be divided equally into four sides of a square.

or or

(D) Four short sticks cannot be divided into two equal parts to complete two other sides of a square.

(E) Six long sticks cannot be divided into four equal parts to construct a square.

5 Point Solutions

17 **(B) 7**

The numbers opposite 3 and 1 are 4 and 6, respectively. At this moment the die has 1 on top, 3 on the right face, 4 on the left, 6 at the bottom. As we roll the die five times from its current position to its final position, the number on top face will be, respectively, 1-4-6-3-1-4. Note that the face facing the viewer has a 2 at all stages of the rolling. If the top face has a 4, the number on the right face is 1. The visible faces when we stop are 4 (top), 1 (right), and 2 (facing us), and their sum is 4 + 1 + 2 = 7.

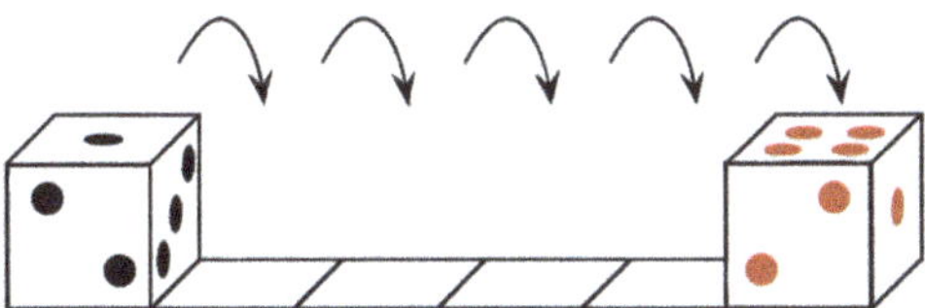

18 **(C) lemon with a wafer**

There are three cherries, so they must be assigned to one vanilla, one chocolate, and one lemon scoop in order to avoid two identical combinations. There is only one lemon scoop that was topped with a cherry, so lemon cannot be topped with a wafer, which is the combination that cannot occur, listed in answer (C). All other combinations mentioned in all answers but (C) are possible. This table shows the allowable possible distribution of ice cream desserts.

	vanilla	vanilla	vanilla	chocolate	chocolate	lemon
3 cherries	x			x		x
2 wafers		x			x	
1 choc. chip			x			

19 **(A) Abbey Lilly Cora**

One way to see the correct names is to test the given five answers against the three statements of the problem and to notice that only (A) satisfies all conditions. Another way is to list all possible combinations of names and test them against the clues given in the three statements. There are two possible first names: Adele and Abbey, two possible middle names: Lilly and Laura, and two possible third names: Cleo and Cora. There are 2 × 2 × 2 = 8 possible combinations for the names:

1. Adele Lilly Cleo
2. Adele Lilly Cora
3. Adele Laura Cleo
4. Adele Laura Cora
5. Abbey Lilly Cleo
6. **Abbey Lilly Cora**
7. Abbey Laura Cleo
8. Abbey Laura Cora

Testing each against the given statements, only combination number 6 makes them correct. So the correct combination of names is Abbey Lilly Cora.

20 **(D) 6**

The sum of all eight numbers is 1 + 2 + ... + 8 = 36. The sum of those covered by squares and triangles is 10 + 20 = 30. So, the remaining number is 36 – 30 = 6.

21 (D) 5

Jane has three colors to use and three body parts to color. Once she colors the first part one of the three colors (for example, coloring the head red), she can color the other two part in only two ways (in this example, either the wings will be green and the tail will be blue, or the wings will be blue and the tail green). So, there are 3 × 2 = 6 different color combinations. Since Jane already colored the parrot one way, she can color 5 more drawings differently.

22 (D) 8

Notice that the number of teams must be less than 9. If there were more than 9 teams, then the number of members would be at least 9 × 5 = 45, and we know there are 43. Additionally, the number of teams must be more than 7. If there were 7 or less teams, then the number of members would be at most 7 × 6 = 42 and we know there are 43. In other words, the number of teams is less than 9 but more than 7, which can only be 8.

Since there are 43 members, there are 5 teams with 5 members each and 3 teams with 6 members each. This makes 5 + 3 = 8 teams and (5 × 5) + (3 × 6) = 43 members.

23 (B)

Figures (A), (C), (D), and (E) can each be cut into three different parts of five squares as shown below in different colors.

Figure B cannot. At the beginning we start with the bottom part of five pieces that must be as shown (blue color). The remaining part (red color) is symmetrical and can only be cut into two identical pieces. One example is the division indicated with white marks on red. It can be verified by trial and error that all divisions would give two identical parts.

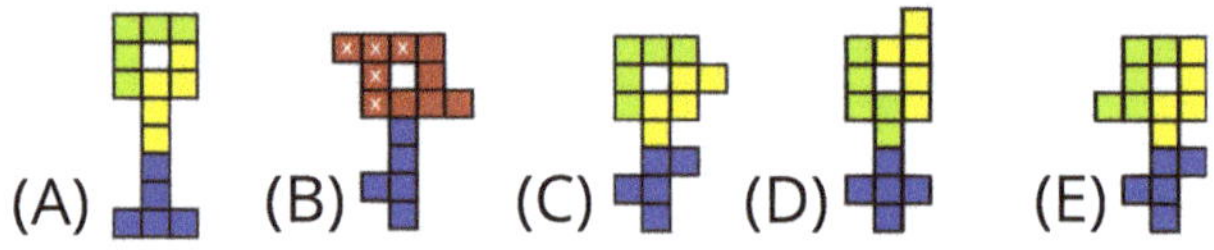

24 (D) 942

In order to achieve the largest possible result, the number ROO that is subtracted should be as small as possible, which makes ROO = 122. In the numbers KAN and GA which are added together, the only hundreds digit is K, so it should be the largest possible, which results in K = 9. The next largest numbers are the tens digits A and G so they should be either A = 8 and G = 7 or A = 7 and G = 8. It can be decided later which one is the correct one. Next, there are two ones digits left: N and A. The second one is A for which at this stage there are two choices. The largest available number left should be chosen for N, which results in N = 6. Using these conclusions we have two options:

KAN + GA = 986 + 78 = 1064 and
KAN + GA = 976 + 87 = 1063.

The first option gives a larger result.
The final possible result that Ann can get is expressed with
KAN + GA – ROO = 986 + 78 – 122 = 942.

2022

3 Point Solutions

1 **(A) → ↓ → ↓ ↓ →**
Buzz needs to make three moves to the right and three moves down in any order to reach the flower. Option (A) lists these six moves (right, down, right, down, down, right).

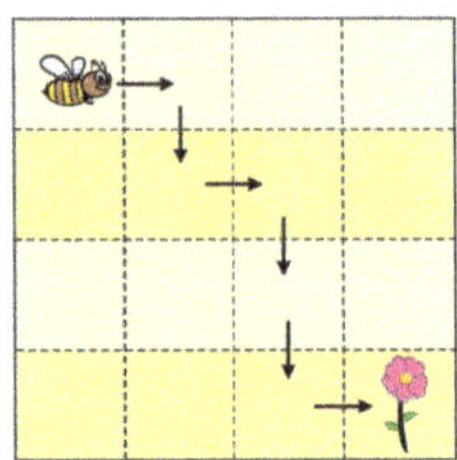

Here is where Buzz would end up by following the other sets of directions:

(B)
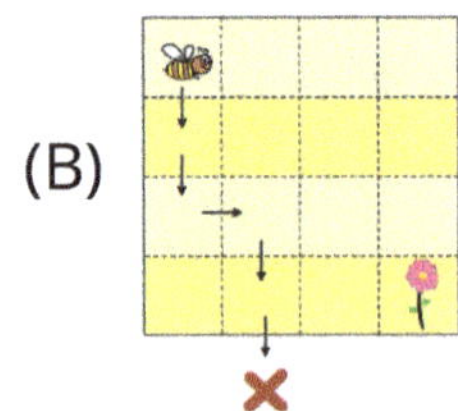

(C)
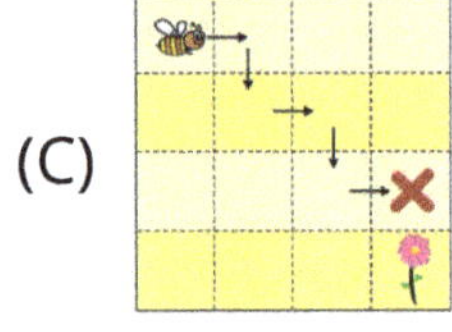

(D)
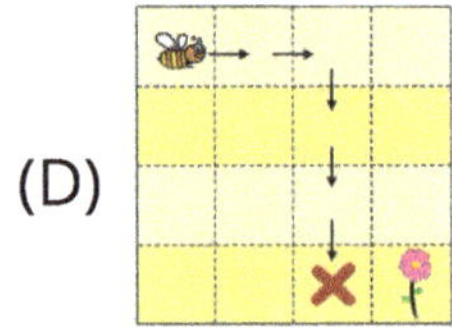

(E)
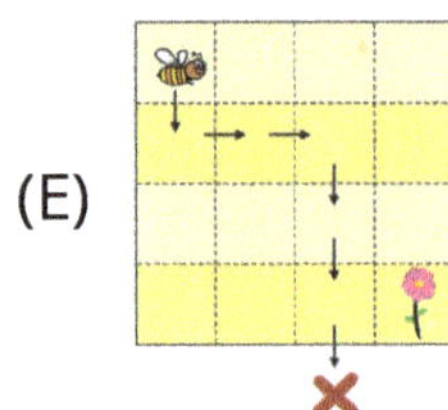

2 **(B)** ***B***
The laser beam travels down, then right, up, right, down, left, down, and ends up at point *B*.

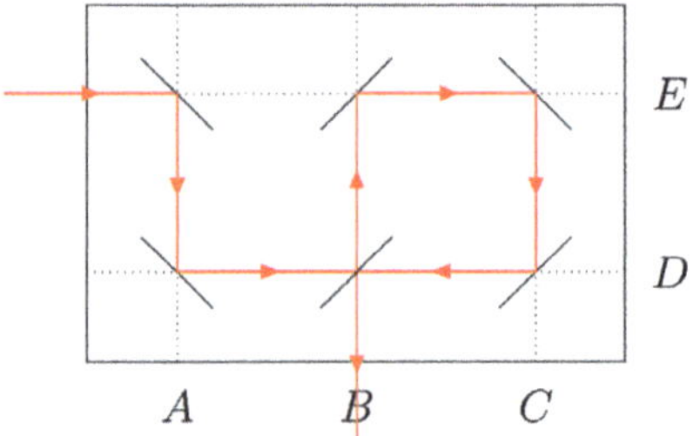

3 **(C) C**
Most rows and columns have two coins, but the first column and the last row have 3 coins each. Coin C is in both the first column and the last row, so it is the one that needs to be moved. (It should be moved to a row and a column that only have one coin each, which are the second column and the fourth row.)

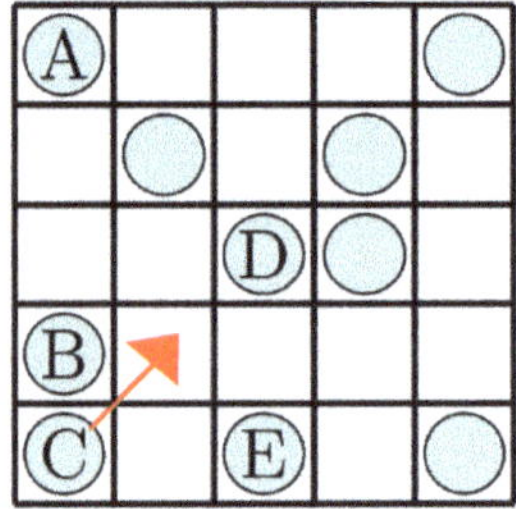

4 **(C) 5**
There are two boxes directly on top of the TRAIN box: stuffed animals and bedding. There are three boxes on top of those: books on top of stuffed animals, board games on top of bedding, and sheet music on top of both. These 5 boxes all need to be moved in order to open the TRAIN box.

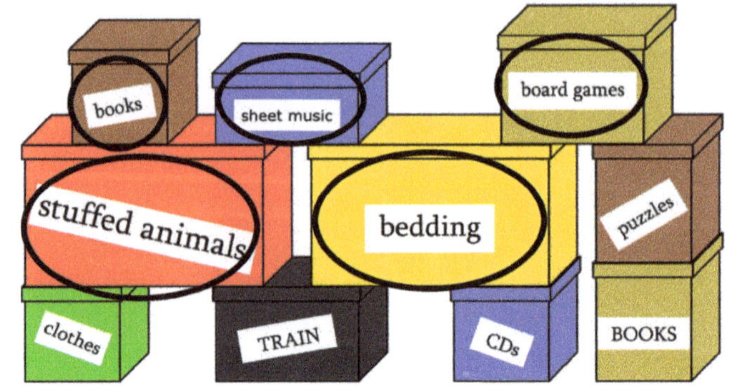

5 **(E) 12**
The sum of the distance of one large jump and two small jumps is 4. This pattern is repeated. 16 is divisible by 4. 16 ÷ 4 = 4 so Kengu repeats the pattern (large, small, small) 4 times. In this way, Kengu makes 4 × 3 = 12 jumps.

---

6 **(D)**

2		
3	1	4

The neighboring squares for the spot marked with the red circle have the numbers 2, 3, and 5. So, the square that should be placed where the red circle is has to have the number 1 or 4. There is no 1 in the far right square of any of the pieces shown among the options, so the number will be 4. There are two pieces with a 4 there, (D) and (E). The neighboring squares for the spot with the green circle have the numbers 3 and 5, so that square can have 1, 2, or 4. Piece (E) has a 3, so it does not fit the puzzle. Piece (D) has a 2 in the upper square. Its other two squares with 3 and 1 also do not have the same numbers as neighboring squares, so only piece (D) fits the puzzle.

3	2	5	4	2	1
1	4	3	1	3	4
2	5	●	5	2	1
4	1			●	3
3	2	4	2	5	2
4	1	3	1	3	4

---

7 **(A) 3 and 5**
Notice that the number 2022 is larger than 2020 by 2. So, in order for the equation to be correct the number in the first square needs to be smaller than the number in the second square by 2. Among the pairs of numbers listed, only 3 and 5 differ by 2. 3 is 2 less than 5, so they can be placed in the squares. 2022 + 3 = 2020 + 5 = 2025

8 **(C)**

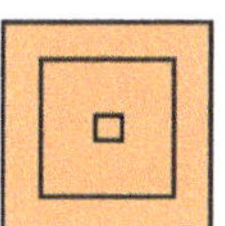

In the picture, five solid figures of different sizes are stacked. When looking down at these solid figures from above, the third and fourth figures are invisible because they are smaller than the second figure.

4 Point Solutions

9 **(B) 2, 1, 3, 5, 4**
The order of the cars changes like this:

1, 2, 3, 4, 5 → 1, 2, 5, 3, 4 → 1, 3, 2, 5, 4 → 2, 1, 3, 5, 4.

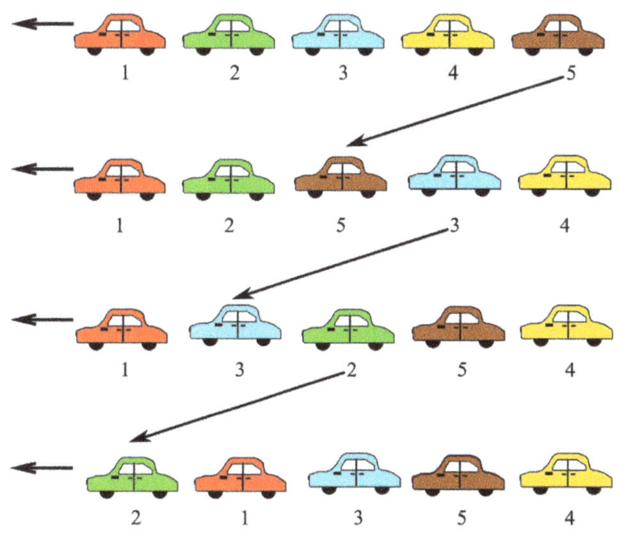

---

10 **(C) 5 and 8**
The sum of all the ages is 2 + 4 + 5 + 6 + 8 + 10 = 35, so the other two ages add up to 35 − 22 = 13. In order for two numbers to make an odd sum, one of the numbers needs to be odd. There is only one odd number among the ages, 5, so this is the age of one of the kangaroos. The other age is 13 − 5 = 8.

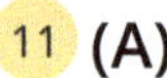

11 (A)

Let's start with the clues in reverse order:

There are kangaroos on Heather's card, so her card must be B.

Lexi's card has a dog on it, so it's E.

There are exactly two living creatures on Paula's card, but we know that B is Heather's, so C is Paula's.

Cara's card has a sun on it, so it could be A or D. However, these are the only cards left, and Mike's card does not have ducks on it.

Therefore, Cara's card is D and Mike's card must be A.

---

12 (B) 3

The sum in most rows and columns is 15. In the first column and the second row, the sum is 16. So, George must change the number that is in both the first column and the second row, which is the 3. (He needs to make it 1 less.)

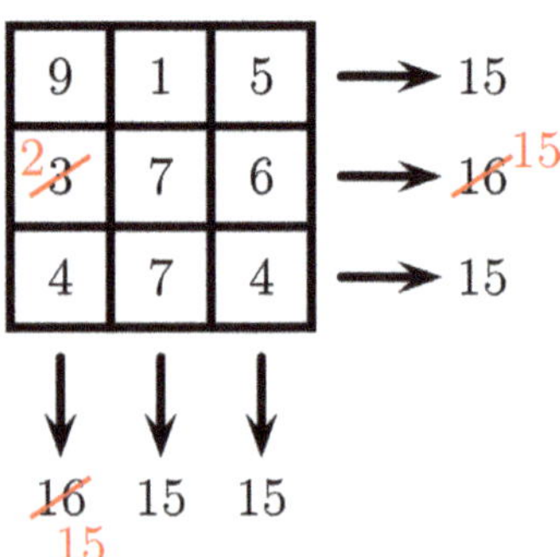

13 (E) 32

The same number of dots arranged in two rows means there are two rows of six dots along each side. The unfolded square carpet will look as shown here. You can now count the dots: 4 rows with 6 dots in each of them and 2 rows with 4 dots in each of these rows make $(4 \times 6) + (2 \times 4) = 24 + 8 = 32$ dots.

You can also notice that if the whole carpet was covered with dots, it would have $6 \times 6 = 36$ dots. Two rows of two dots (4 dots) are missing in the middle. So, there are $36 - 4 = 32$ dots.

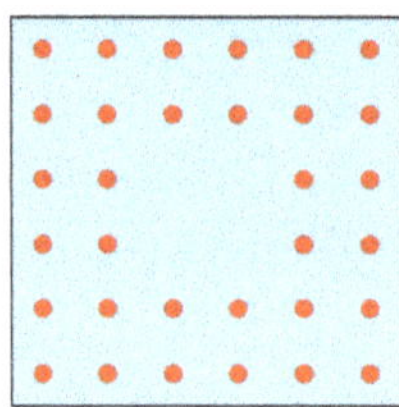

---

14 (B) 14, 17, 20, 23

The picture shows the paper and the holes as we unfold it to its original shape. The numbers Joanna punched through are marked.

1	2	3	4	5	6
7	8	9	10	11	12
13	14	15	16	17	18
19	20	21	22	23	24
25	26	27	28	29	30
31	32	33	34	35	36

---

15 (E) 36

Since there is the same number of students in each row, the students sitting in class form a rectangle. There are 2 students in front of Robert and one behind him. This indicates the number of rows, so there are 2 + 1 (Robert) + 1 = 4 rows. In Robert's row, there are 3 students on one side of him and 5 on the other side, so there are 3 + 1 (Robert) + 5 = 9 students in each row. Therefore, the total number of students is $4 \times 9 = 36$.

16 **(B) 11**

The large cube can be made using a total of 3 × 3 × 3 = 27 small white blocks (cubes). The light gray blocks can be formed using 5 white blocks, and dark gray blocks can be formed using 3 white blocks. In the picture, you can see that 2 light gray blocks (one fully visible, and the other with ends at the very top and at the bottom to the right). You can also see 2 dark gray blocks (one fully visible, and the end of another one). There is no place in the cube to fit any more of the light or dark gray blocks. Since a completely white cube would use 27 small white blocks in total, we can subtract the space the gray blocks are taking up. This cube uses 27−10 (2 light gray blocks)−6 (2 dark gray blocks) = 11 white blocks were used.

5 Point Solutions

17 **(B) 3**

Wanda had to choose the two circles to have 2 round shapes, and so she already has one large shape and one colored shape. She then can choose the other large colored shape, the square, to have 2 large shapes and 2 colored shapes. So, the smallest number of shapes that Wanda could have chosen is three.

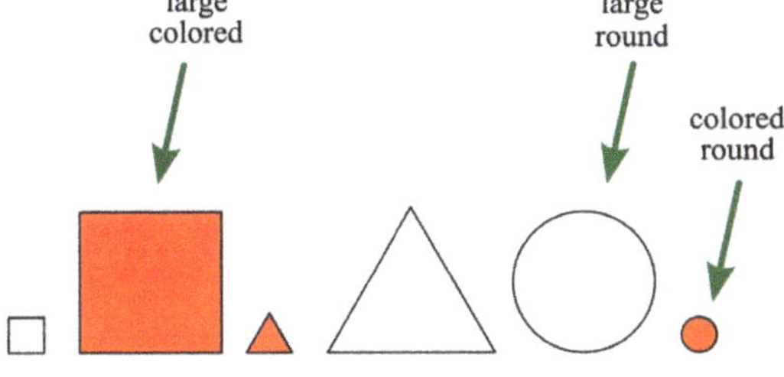

18 **(D) 5**

Each team has just 2 games and the possibility to get 0, 1, or 3 points in each game, so the only possible numbers of points for any team to have at the end of the tournament are 0 + 0 = 0, 0 + 1 = 1, 1 + 1 = 2, 1 + 3 = 4, and 3 + 3 = 6. To get 5 points, a team would have to play at least 3 games (3 + 1 + 1 points).

19 **(E) 90 cm**

The ant walked up three segments equal to the height of one of the cubes and down three such segments, so its path had six vertical segments. Their length is 6 × 10 cm = 60 cm. In the view from above we can see that the horizontal part of the ant's path is equal to three of the cube's edges, so its length is 3 × 10 cm = 30 cm. In total, the ant's path is 60 cm + 30 cm = 90 cm.

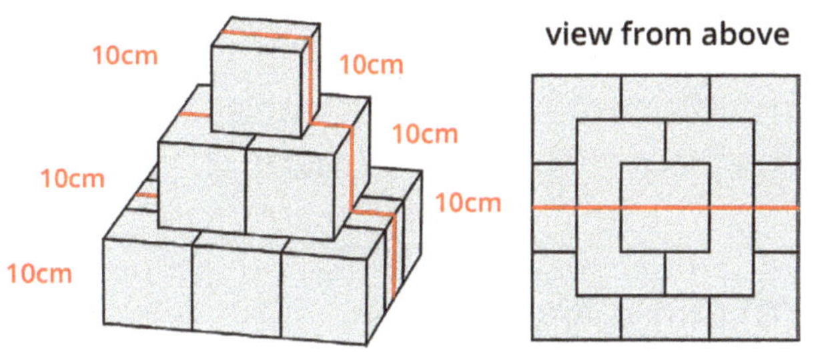

20 **(E) 1 and 5**

Since E and B are across from each other, the pieces need to have a straight road so these two houses are connected to the middle. A is counterclockwise from B, so there also needs to be a road just counterclockwise from the straight segment. However, that road cannot continue straight, because D cannot be connected. The tile below shows in black the roads that need to be there, and in red the one that cannot. Any other road segments on the tile don't matter.

The pictures below show the pieces rotated so that one of the paths starts at A. Only pieces 1 and 5 have the required roads to B and E without having a road to D.

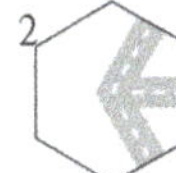

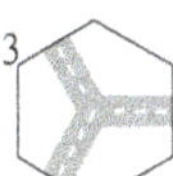

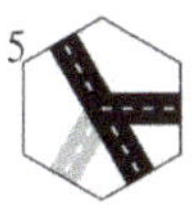

21 (C) 3

Each of Ahmad's laps is 4 × 5 m = 20 m long, and each of Zhaleh's laps is 2 × 5 m + 2 × 10 m = 30 m long. They can only meet at A when each has completed full laps and their total distance is the same. Ahmad completes full laps after walking 20 m, 40 m, 60 m, 80 m, 100 m, and so on. Zhaleh completes full laps after walking 30 m, 60 m, 90 m, 120 m, 150 m, and so on. The smallest number found in both lists (the least common multiple of 20 and 30) is 60 m. After walking 60 m, Ahmad has made 3 laps and Zhaleh has made 2.

22 (B) Claire and Sophie

To solve the problem, it is helpful to mark the numbers of plums on a number line. Lauren ate 2 plums more than Sophie. Because Betty ate 3 fewer plums than Lauren, Betty ate 1 plum fewer than Sophie. Since Claire ate 1 plum more than Betty, Claire ate as many plums as Sophie. So, (B) is correct. Alice ate 3 plums more than Claire, so Alice ate 1 plum more than Lauren. All the numbers are different except for Claire and Sophie.

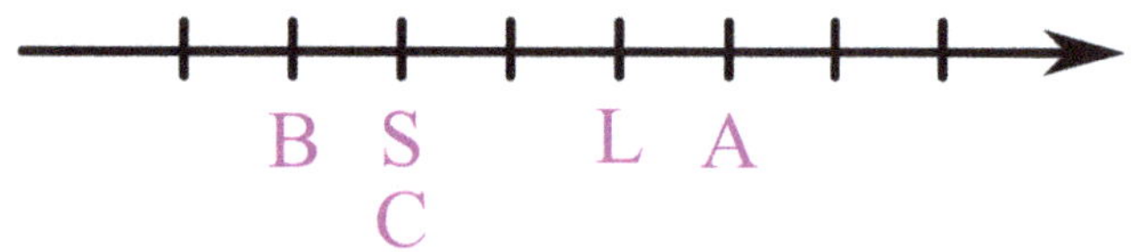

23 (A) 

Although it does not matter which end we start at, we can assume that the head of the caterpillar is the orange circle. The body has an alternating pattern of one yellow and one black circle. You have to find a possible "path" through the sleeping caterpillar where you can draw one line from the head to a black circle, then to a yellow circle, then to black, and so on. The only possibility to find such a way through whole the body of the caterpillar is in option A, as shown:

24 (D) 12

Looking at the first row, we see that the number hidden under a gray square plus two numbers hidden under white squares add up to 34. Looking at the third row, we see that the number hidden under a white square plus two numbers hidden under gray squares add up to 26. This means that three numbers under gray squares plus three numbers under white squares add up 34 + 26 = 60. Thus, the number under a white square plus the number under a gray square equals 60 ÷ 3 = 20.

If we look at the first row and the last row, we also see that the difference between the two row sums is 34 − 26 = 8. By removing one white and one gray square from each of the two rows we keep the difference the same, so "white" − "gray" = 8. The only two numbers for which the sum is 20 and the difference is 8 are 14 and 6 (14 + 6 = 20 and 14 − 6 = 8), so "white" = 14 and "gray" = 6. In the second row, "gray" + "white" + "black" = 32. "gray" + "white" = 20, so "black" = 32 − 20 = 12.

Part III
Answer Key

	1998	2000	2002	2004	2006	2008
1	C	B	B	E	C	C
2	D	C	C	A	B	D
3	C	D	D	C	D	B
4	D	B	D	C	A	B
5	C	B	D	E	D	D
6	C	C	C	B	E	A
7	D	B	A	E	B	E
8	B	C	B	B	B	A
9	C	B	C	D	C	E
10	E	D	E	D	A	D
11	C	B	C	B	A	E
12	C	D	A	B	D	D
13	A	B	E	E	C	E
14	B	C	B	B	E	C
15	D	C	C	E	A	B
16	E	C	E	C	C	B
17	B	B	A	E	B	C
18	D	B	B	C	D	A
19	E	E	D	C	D	D
20	D	C	E	A	E	E
21	B	C	C	A	B	E
22	B	A	D	B	E	B
23	B	E	E	E	C	C
24	D	A	D	E	B	D

	2010	2012	2014	2016	2018	2020	2022
1	D	B	D	E	B	E	A
2	C	D	D	E	E	E	B
3	C	B	A	A	C	A	C
4	C	C	D	A	E	A	C
5	D	A	A	D	D	E	E
6	A	E	E	B	D	D	D
7	C	B	E	A	D	E	A
8	B	D	C	C	A	C	C
9	E	E	E	B	D	C	B
10	B	B	E	B	A	C	C
11	B	E	B	C	B	D	A
12	D	C	D	B	E	E	B
13	D	D	B	D	E	D	E
14	C	B	B	A	C	A	B
15	D	D	C	C	C	D	E
16	A	C	B	D	D	B	B
17	E	D	B	D	D	B	B
18	C	C	C	B	C	C	D
19	D	D	B	D	B	A	E
20	D	B	C	C	A	D	E
21	E	E	A	B	B	D	C
22	B	D	D	C	D	D	B
23	D	E	D	E	E	B	A
24	E	C	D	B	B	D	D

www.ingramcontent.com/pod-product-compliance
Lightning Source LLC
LaVergne TN
LVHW060639110826
845147LV00018B/1008
* 9 7 9 8 9 8 9 9 8 8 3 0 3 *